The Whole Creature

The Whole Creature

Complexity, biosemiotics and the evolution of culture

WENDY WHEELER

Lawrence & Wishart
LONDON 2006

© Wendy Wheeler 2006

All rights reserved. Apart from any fair dealing for the purpose of private study, research, criticism or review, no part of this publication may be reproduced, stored in a retrieval system, or transmitted, in any form or by any means, electronic, electrical, chemical, mechanical, optical, photocopying, recording or otherwise, without the prior permission of the copyright holder.

The author has asserted her rights under the Copyright, Design and Patents Act, 1998 to be identified as the author of this work.

British Library Cataloguing in Publication Data.
A catalogue record for this book is available from the British Library
ISBN 978-1-905007-30-1

Printed in Great Britain by Biddles, Kings Lynn

In memory of my dear son
Ollie Thomas Wheeler
25th September 1979–3rd August 2003

*Para Carmencita
en el día de su
cumpleaños 2008*

You need not see what someone is doing
to know if it is his vocation,

you only have to watch his eyes:
a cook mixing a sauce, a surgeon

making a primary incision,
a clerk completing a bill of lading,

wear the same rapt expression,
forgetting themselves in a function.

How beautiful it is,
that eye-on-the-object look.

To ignore the appetitive goddesses,
to desert the formidable shrines

of Rhea, Aphrodite, Demeter, Diana,
to pray instead to St Phocas,

St Barbara, San Saturnino,
or whoever one's patron is,

that one may be worthy of their mystery,
what a prodigious step to have taken.

There should be monuments, there should be odes,
to the nameless heroes who took it first,

to the first flaker of flints
who forgot his dinner,

the first collector of sea-shells
to remain celibate.

Where should we be but for them?
Feral still, un-housetrained, still

wandering through forests without
a consonant to our names,

slaves of Dame Kind, lacking
all notion of a city,

and, at this noon, for this death,
there would be no agents.

W.H. Auden, Horae Canonicae

Contents

Acknowledgements	9
Introduction A very long revolution	12
Chapter 1 A long time coming: the complexity revolution	38
Chapter 2 'That eye-on-the-object look': complex culture and the passionate structure of tacit knowledge	60
Chapter 3 Intimations: nested enfoldings – stages in knowing	80
Chapter 4 Perfused with signs: biosemiotics and human sociality	106
Chapter 5 The importance of creativity	131
Bibliography	161
Index	167

Acknowledgements

This book has been hatching since 1998 at least, and, in some form, perhaps much longer. In 2002, a sabbatical from London Metropolitan University enabled me to get started on the task of sorting out my ideas. Creativity, as *Complexity, Biosemiotics and the Evolution of Culture* argues, is always a collective affair. Many colleagues and friends have helped me in this process, not only by discussions, encouragement to give early versions at various conferences and symposia, and reading and commentary on early drafts, but also in offering the general support which kept me at the task of reading and writing. In particular, I would like to thank Roger Luckhurst for being a stalwart and thoughtful friend during an immensely difficult time in my personal life, and for deviously manoeuvring an invitation to me to give a paper, at King's College London in the Spring of 2004, when I thought I might have run out of words for anything at all. It was a good trick which, despite my initial hesitancy, worked well. There were some things about human life, biology and mind that I needed to be able to think about in a different way; the paper for King's made me start on this properly. Sometimes the shocks of a life can render you speechless; sometimes being obliged to open your mouth (even before your brain seems fully engaged) can produce words, like children, with a life of their own. In June of the same year, when I was trying to mark essays, this book started to write itself; it felt like a welcome, if untimely, labour. Readers will judge for themselves; but, for me, I think it was as much a labour of love as of mind.

It seems entirely appropriate that a book about human sociality should be born out of the experience of much friendly support. My second debt is to another friend, Jonathan Rutherford. In the Winter of 2003-04, I mentioned to Jonathan that my interest in complexity theory, and the origins of social totality thinking in German Romanticism, had led me to some re-readings of Engels and Marx. Jonathan asked me to write an article on this for the journal *Soundings*. Following the initial 'unlocking' of May 2004, this article, 'The Complexity Revolution', published in the Spring 2005 issue of *Soundings* entitled 'After Identity', formed the beginning of the June 2004 thaw which helped me to shape up the ideas which now form chapter 1 of the present work. In the Spring of 2005, Jonathan read through the whole first draft of the manuscript and offered some particularly astute comments which led me to a much more focused re-writing of chapter 5. For this I am grateful.

Yet again, re-engaging in Autumn 2004 with the research life of the University in the newly expanded area of Communications, Cultural and Media Studies at London Metropolitan University – formed by the merger between London Guildhall University and University of North London in 2002 – I was fortunate enough to meet and work with a new colleague from the old Guildhall University, Paul Cobley. Paul's research liveliness – and especially his interest in biosemiotics, which he pursued via symposia he organised at the University, and where I was able to meet and talk with leading biosemioticians Jesper Hoffmeyer, Frederik Stjernfelt, Soren Brier and Kalevi Kull in Spring 2005 – were absolutely central to solving a problem I had encountered, in writing the book, of finding a new language in which the epidemiological and psychoneuroimmunological research I had been led to could be expressed in ways which were meaningful both scientifically and aesthetically. Paul introduced me to biosemiotics, and has also helped me tremendously by reading and commenting upon the whole draft manuscript of this book. I now count Paul not only a new colleague, but a new friend too.

Other much valued friends and colleagues who have read and commented upon early drafts include Steve Baker and Carolyn Burdett. Steve's willingness to run creatively with my ideas especially as those might be worked through in his own field of Art History and animal studies have proved consistently enlightening. This has contributed to a continuing and deeply fruitful conversation which Steve and I have been holding for more than a decade now. I am grateful for his reliability and persistence as both friend and intellectual partner in crime. Carolyn has been, as ever, a constant friend and interlocutor. Her acuity, affection, and maturity about human relationships, and her continual thoughtfulness about the opacity of conscious and unconscious communication, has had incalculably transforming effects upon my thinking for many years. Without the on-going and often daily dialogue I have enjoyed with her, I could not have developed the sense, and sensibility, about human communication and its strangeness which has made thinking through the matters explored in this book possible. To Carolyn thanks, too, for the wholly apposite verse from Auden on tacit skills which is now the book's epigraph.

To the discovery of the epidemiological and related sociological work I discuss here, I am entirely indebted to another friend and colleague at London Metropolitan University: Helen Crowley. As with so much which appears serendipitous in the creative engagement with ideas, this help sprang, appropriately, from a very womanly conversation about the experiential and practical aspects of life child-care in relation to teaching commitments in this case. In the course of this, an

Acknowledgements

enquiry about the progress of my research led to Helen's advice that I look at what Michael Marmot and Robert D. Putnam had to say. This, too, was invaluable in helping me move forward with the arguments I needed to make.

The librarians at London Metropolitan University have been constantly helpful, as have the postgraduate students on the MA in Literature and Modernity in whose company some of the ideas contained in this book were initially articulated and explored. I am also grateful to Phil Tew, whose offer of support from the London Modern Fiction Network for a conference on any subject I cared to pursue led to the May 2000 conference on complexity theory at the University of North London. This, in turn, led to the special issue of *New Formations* ('Complex Figures', *New Formations* 49, Spring 2003). In co-editing this with Phil, I was able to develop my understandings of complexity theory further. I am, consequently, also indebted to *New Formations* editor, Scott McCracken, and to the Editorial Board of *New Formations* more generally, for affording me the opportunity of co-producing the special issue.

Special thanks must go, also, to my editor at Lawrence & Wishart, Sally Davison, who has always been both appropriately challenging and entirely supportive. I am very grateful not only for her keen editorial eye, but also for her encouraging receptivity to some new ways of thinking about the old themes to which both she and I are so passionately committed.

Lastly, my love and thanks go to my daughters, Bella, Beattie and Tilly, whose toleration of the endless hours and sometimes maternal absent-mindedness involved in the hard work of researching and writing a book has been generally (regular objections to the late appearance of supper notwithstanding) both loving and generous. In dedicating this book to the memory of their brother Ollie whom all of us miss dreadfully I do not, of course, forget the hard work of various kinds which they, too, have had to do.

Wendy Wheeler, London, November 2005

INTRODUCTION

A very long revolution

> In naming the great process of change the long revolution, I am trying to learn to assent to it, an adequate assent of mind and spirit. I find increasingly that the values and meanings I need are all in this process of change. If it is pointed out, in traditional terms, that democracy, industry, and extended communications are all means rather than ends, I reply that this, precisely, is their revolutionary character, and that to realise and accept this requires new ways of thinking and feeling, new conceptions of relationships, which we must try to explore. This book is a record of such an attempt.
> Raymond Williams, *The Long Revolution*[1]

In this book, and following in an honourable tradition which I have found best expressed in the work of Raymond Williams, I have wanted to present a good materialist argument about the nature of human sociality. My argument, briefly stated, is simple: while each and every one of us is manifestly an individual, whose life and wellbeing matters, humans and their wellbeing are not most fully understood unless the fundamentally social nature of human existence is properly taken into account. This – our fundamental sociality – is lived in our inner, as well as outer, world; and it is emotional as well as physical; and all this – our essential social being – is written on our bodies in terms of flourishing or (its opposite) illness. But believing this as a common observation is not the same as making a good, and amongst other things, scientifically based argument for how it is so. The purpose of this book, therefore, is to elucidate this opening premise by bringing together some of the most relevant arguments, across various disciplines, including the sciences, in order, hopefully, to persuade the reader of the truth of its claim.

This endeavour will involve, first, an encounter with the long revolution in human thought and scientific understanding which has gradually been building in the twentieth, and now twenty-first, centuries: the revolution that is taking place in the scientific paradigm shift from 'The Age of Reduction' to 'The Age of Emergence'. Second, it will involve a consideration of the evolutionary processes, in nature and in culture, by the understanding of which humans began to grasp the complex structures through which life itself – and human knowing of it – are accomplished.

By drawing on developments in the sciences, particularly complexity science, I hope to show that sociality can be seen as firmly rooted in an account of evolution that sees it as a process of symbiogenetic co-operative communication (from the cell all the way up), with the consequent emergence of more complex levels of life.[2] Williams's remarkable perspicacity lay in catching sight of this long revolution, and its implications, so fully and so quickly. I use the term 'long revolution' in acknowledgement of Williams's perception of the processes of cultural change in modernity. But I am particularly referring to a now quite distinct transition – which he certainly perceived in outline – from an age of scientific reductionism to an understanding of the importance of complexity and emergence. It has been objected that this transition does not represent a paradigm shift in our understandings, but is simply a normal turn in the development of science. However, I would argue that the transition to a complexity/emergence understanding of change potentially transforms so much in our customary modern understandings of the world that it does deserve to be called a shift in paradigmatic thought in Western modernity. In this book I look at some of its most striking implications. Most importantly, my argument is that recent developments in science confirm that there is a material basis for human sociality, and that the shift I describe allows us to see more clearly the relationship between individual, culture and society.

Thus this book grapples with science as well as culture. For an attempt to understand human behaviour, the most obvious engagement with science that the humanities and human sciences could make is with the biological sciences. But until the second half of the twentieth century, biology remained almost entirely caught within a reductionist framework, thus limiting its usefulness for the kinds of understanding sought in this book. Meanwhile, the human sciences were able to benefit to some extent from the use of scientific methodologies. But the full complexity of human behaviour and motivation could never be satisfactorily caught by such reductionist means. The humanities – i.e. the study of human complexity – barely bothered with biology at all. Williams was unusual for his time in drawing upon biology and evolution for his account of the long revolution. His own engagement with complexity (*avant la lettre*) and biology – both evolution and the biology of perception – in *The Long Revolution* eschewed any 'naïve realism'. His essentially 'ecological' understanding rejected both positivism's sharp distinction between subject and object, and also its reduction of either to a simple machinery. For an understanding of a complex biology of perception he turned to the marxist Christopher

Caudwell's flawed but nonetheless far-sighted account: '"Body and environment are in constant determining relations. Perception is not the decoding of tappings on the skin. It is a determining relation between neural and environmental electrons. Every part of the body not only affects the other parts but is also in determining relations with the rest of reality. It is determined by it and determines it..."'.[3] Williams also understood the importance of communication in nature as well as in culture. But, if his attempt to think a complex non-reductive and non-positivistic semiotic biology and culture together came a decade before the main onslaught of a determinedly reductionist sociobiology and neo-Darwinism, it also came before the equally distracting – although not so violently misguided – arrival of Saussurean semiology in anglophone cultures. *The Whole Creature* once more asserts the need for a return to biology – to break the great nervous silence in progressive thought on culture – in order to take up the task which Williams began nearly fifty years ago.

One reason this is now both timely and possible is that recent developments in molecular and developmental biology are radically altering our understanding of evolutionary biology as formulated by Darwin. As Eva Jablonka and Marion J. Lamb point out in their 2005 book *Evolution in Four Dimensions: Genetic, Epigenetic, Behavioral, and Symbolic Variation in the History of Life*,[4] it is now clear that the gene-centred view of evolution via which neo-Darwinists have attempted to reduce the complexity of human life to a genetic determinism is seriously incomplete. As they write,

> Molecular biology has shown that many of the old assumptions about the genetic system, which is the basis of present-day neo-Darwinian theory, are incorrect. It has also shown that cells can transmit information to daughter cells through non-DNA (epigenetic) inheritance. This means that all organisms have at least two systems of heredity. In addition, many animals transmit information to others by behavioural means, which gives them a third heredity system. And we humans have a fourth, because symbol-based inheritance, particularly language, plays a substantial role in our evolution. It is therefore quite wrong to think about heredity and evolution solely in terms of the genetic system. Epigenetic, behavioural, and symbolic inheritance also provide variation on which natural selection can act.[5]

Not only has the inheritance of acquired characteristics been shown to be possible,[6] but such epigenetic inheritance indicates our inseparable lived relation to our environment, including our cultural environment. And if what we feel and experience both biologically and psychobiologically is capable of heritability, this has very far-reaching social, ethical and political implications.

Early drafting of Williams's *The Long Revolution* (1961) began in the late 1950s, before *Culture and Society* (1958) was even published. The latter was conceived as 'an account and interpretation of our responses in thought and feeling to changes in English society since the later eighteenth century',[7] and *The Long Revolution* 'was planned and written as a continuation' to it.[8] In it, Williams attempted to imagine that long revolution more intensely in order to think both where it might be leading, and how (the intellectual, including scientific, means by which) it might be assented to. His first chapter, 'The Creative Mind', covers all the material, in necessarily embryonic form, which I have found myself wishing to attend to – with the benefit of a further nearly fifty years development in human thought – in more detail in the present work. The logic of the movement is now more obvious than it might have seemed to some readers of *The Long Revolution* forty-five years ago, but I am still impressed by, and indebted to, Williams's grasp of the things it would be important to understand, and to understand more about, in order to approach in some detail the questions about the ways in which the long revolution of modernity might unfold.

There is a range of concerns expressed in 'The Creative Mind' – human creativity; the biology of perception and the evolved relation between (human) creature and environment; the need for an understanding of complex systems; the semiotic nature of *all* life; and, above all, the need for a materialist, but non-positivistic and non-reductionist, account of evolutionary cultural change – that can be seen as nascent forms of these later developments. Particularly important from my point of view was Williams's insight, derived from then contemporary developments in the biology of perception, of the ways in which human species perception is both embodied and enworlded. The world we sense is not there in any positivistic 'objective' sense, but is made by evolution in the conjunction of creature and environment. For humans, environment includes, of course, culture – and Williams was very conscious of the way in which this, too, is fundamental to creative human evolutionary change. Because of its prescience in relation to biological and social complexity, it's worth quoting the relevant passage at length:

> Thus man shares with all living creatures this fundamental process, but in fact has evolved in such a way that his 'building of organisation' is a continual process of learning and relearning, as compared with the relatively fixed instinct mechanisms of animals. It is man's nature, and the history of his evolution, to be continually learning by the processes described [binding, dissolution, and rebinding, or organisation, disorganisation and reorganisation]. Since this continuing organisation and reorganisation of consciousness is, for man, the organisation and reorganisation of reality – the consciousness a way of learning to control his environment – it is clear that there is a real sense in which man can be called a creator.

All living forms have communication systems of a kind, but again, in man, the process of learning and relearning, which is made possible by social organisation and tradition, has led to a number of communication-systems of great complexity and power. Gesture, language, music, mathematics are all systems of this kind. We can think of them as separate systems, yet to understand their nature, in any depth, we must see them in their context of the whole process of social learning. At one level we can oppose art to science, or emotion to reason, yet the activities described by these names are in fact deeply related parts of the whole human process. We cannot refer science to the object and art to the subject, for the view of human activity we are seeking to grasp rejects this duality of subject and object: the consciousness is part of the reality, and the reality is part of the consciousness, in the whole process of our living organisation.[9]

THE LINGUISTIC TURN AND NEOLIBERALISM

In some ways, the direction of Williams's thought as represented in *Culture and Society* and *The Long Revolution* can be seen to have been deflected by subsequent intellectual development in the 1960s. In particular, under the considerable influence of Perry Anderson's editorship of *New Left Review*, there was an opening up of untheoretical and complacently pragmatic and empiricist British bourgeois culture (mirrored in the dominance of Labourism on the organised left in Britain) to Continental ideas. With this, left culturalist thought was passed through both Gramsci and also structuralism and post-structuralism. The latter, in particular, introduced what has often been referred to as 'the linguistic turn' in cultural studies in the late 1970s and 1980s and, to some extent, beyond that into the 1990s. Williams's own sense that 'all living forms have communication systems of a kind', and of the idea of the importance of human semiosis more widely understood as an evolved complexity in which nature, in humans, produces culture, was thus to some extent displaced by the post-structuralist tendency to place emphasis on the defiles (and metaphysical assumptions) of articulate language alone. The postmodern linguistic turn thus inhibited the range of its own insights, in its tendency to confine itself to a focus on Saussurean semiology rather than the wider transdisciplinary (including evolutionary biology) development of semiotics found in the work of writers such as Charles Sanders Peirce and, later, Thomas A. Sebeok.

A useful discussion of the problems of the 'linguistic turn' can be found in philosopher John Deely's *The Impact on Philosophy of Semiotics* and *Four Ages of Understanding: The First Postmodern Survey of Philosophy from Ancient Times to the Turn of the Twenty-First Century*.[10] Whilst this is not the place for any detailed discussion, the general direction of Deely's arguments, especially in the latter book, takes the form of an investigation of the development of semiotics from Augustine of Hippo to that of Aquinas and Duns Scotus. This history of semiotics is displaced by Galileo and Descartes and rediscovered by Peirce. Thus, for Deely, the linguistic turn which, like many, I have referred to as 'postmodern', is, in fact, more accurately conceived of as 'ultramodern'. This is because it maintains a philosophical idealism in spite of its efforts to escape this via the semiological (i.e. Saussurean) route of considering semiosis more or less entirely in terms of articulate language alone, and talking of the materiality of this language as constructing reality:

> There are some today, we have seen, who embrace modern philosophy's culminating doctrine that only the mind's own constructions are properly said to be known, ones who have yet tried to coin and appropriate the phrase 'postmodern' to advertise their stance. But the vain appropriation cannot conceal the stipulation which guarantees that these would-be postmoderns are nothing more than surviving remnants of a dying age, the last of the moderns, in fact, the 'ultramoderns'. The future, in philosophy and in intellectual culture more broadly conceived, belongs rather to semiotics, the clearest positive marker we have of the frontier which makes modernity be to the future of philosophy what Latinity was to philosophy's future in the time of Galileo and Descartes – though this time we will hardly be able to repeat Descartes' mistake of counting history as nothing, as the joint work of Williams and Pencak has perhaps best shown.[11]

The linguistic turn produced some influential interpretations, which were in themselves (as Deely argues) metaphysical ('reality is entirely constructed in language'), and also in some ways complicit with the derealisations and alienations of bourgeois liberal philosophy. Human and natural biology are palpably *not* human constructs, either mastered or made. They *are* powerfully semiotic – as I shall argue in detail in Chapter 4 – but they are not 'constructed in discourse'. This is not to say that words and discourses aren't very powerful. They are: as my discussion of epidemiology and psychoneuroimmunology in Chapter 4 indicates. But they are one aspect of semiotic communication amongst other, unconscious and 'gestural', ones. The focus on abstract conceptual knowledge articulated in written or spoken language tends to obscure or occlude this.

Williams's own seriousness about science, although resolutely non-positivistic and non-reductionist, would still be a target for the hostility to a realist ontology that has been expressed in the last three decades towards science's remaining positivism and reductionism. The latter was, rightly, a target for linguistic constructivists; but the methodologies deployed – the exclusive focus on articulate human discourse alone, and the refusal of *any* material truth about how the world is when offered in scientific terms – could not finally survive.

The value relativism that the sociolinguistic constructivist arguments produced – as Kate Soper, for one, has argued – made it impossible to argue for the superiority of one set of ideas (social justice over self interest, for example) above another.[12] For this reason, among others, it is now necessary to push the pendulum back towards an acknowledgement of the materiality of the world. This is made into an easier task by the fact that, although unevenly, scientific positivism has been for some time under steady attack from scientists themselves, as the understanding of complex systems, and especially biological evolutionary complexity, slowly spread.

At the same time that these 'linguistic turn' critiques were going on in cultural studies and the humanities during the 1970s, liberalism, in the form of neo-liberalism, was mounting a serious counter-reaction, pushing for a reversion to a liberal view of individuals and societies. Consequently, in the last three decades we have been living through a furious resurgence – I think of it as the fury of a certain kind of liberalism in its intensified neo-liberal death-throes – of the now more or less global hegemony of a certain set of ideas about what kinds of creatures human beings really are. These ideas, which are summed up in the philosophy and political economy of liberal possessive individualism, seem to me to be fundamentally wrong. Human beings are not most fully comprehended when they are thought of *primarily* as isolated and monadic self-interested individuals. On the contrary, humans are powerfully social creatures. It is my hope that some of what I have to report in the present book will go some small way to adding to the many other voices seeking to correct these lately resurgent liberal accounts of freedom. Human freedom, as the full expression of human capabilities in any and every individual, is absolutely important – not least in terms of individual flourishing expressed in health and a sense of wholeness. But, as I will argue in this book, it is not best understood simply in terms of possession, property, and the pursuit of self-interest, but rather in terms of what the new field of biosemiotics describes as 'semiotic freedom'. We are *most* free when the lives of our bodyminds – which is to say our lives as phenomenologically whole creatures embodied in an environment which also is really a part of us – are socially and politically recognised.

SCIENCE AS CULTURAL EVOLUTION
Since the seventeenth- and eighteenth-century advent of liberal philosophies in Europe, buttressed by a certain bloodless account of reason in which linguistic abstraction (and possibly the Christian heritage) allowed an extraordinary disregard of the body as the substrate of any experience at all, the main and earliest critical responses to liberal philosophy and political economy have been derived from German Romanticism. While the Marxist version of this

took various shapes within the social sciences, early opposition to the liberal account of what it is to be human was in the main a humanities affair – as it is told, for example, in Raymond Williams's *Culture and Society*.[13] And so it has largely remained, not least because the science to which the philosophy of liberal individualism was midwife in the seventeenth and eighteenth centuries was, itself, powerfully influenced by that philosophy's dichotomous world view. The account of human freedom offered in this book does not rest on the existential freedom imagined in liberal philosophy, however, but upon the complex systems of what Jesper Hoffmeyer calls 'semiotic freedom' – a biosemiotic account of freedom as evolutionary development. This is a freedom which is always constrained by grammar and discourse, and the rules which are a part of human social and cultural making, but which is also always open to the 'rule-breaking' evolutionary emergence of the newer grammars and newer languages in which we recognise human creativity. My argument is that biosemiotics – the study of signs and significance in all living things – offers us precisely such a new language.

This book, taking recent developments in science – especially the biology of complex systems – as its main object, will thus (as should already be clear) give no comfort to the view that science is just one discourse amongst others. Languages, discourses, are not themselves properly understood as simple alternatives to a realist ontology; they themselves have a material history and are evolutionary – as any thoughtful consideration of the human evolution of language out of the non-linguistic proto-culture of our ape forebears must, itself, suggest. Articulate language is an evolutionary accomplishment in which the semiosis that is apparent in all nature achieves a new, and more complex, level of articulation. 'Culture' isn't something utterly new and strange in life on earth; it is the form that the semiotic nature of biological evolution takes with the advent of *Homo sapiens*, which is also the advent of articulate language and abstract conceptual thought. Science, in the modern sense, isn't 'just another discourse'; it is a cultural evolution, of such speed that we are inclined to call it a 'revolution', which marks the new development of a tremendously powerful way to find out more about the real world – our most recent and brilliant kind of artistry yet.

But science, as Thomas Kuhn suggested,[14] and as the paragraph above implies, is powerfully influenced by the paradigms, largely discursive, through which we make sense of our world. Linguistic constructivism was right in this regard: science is subject to the pull of narrative and metaphor in the same way as any other form of human cognisance. It is, we might say, itself double-coded: on the one hand it claims to trump, in its modelling of the world, all other frames; but, on the other hand, it remains to a significant extent subject to the frames of its own time and place: these subsist, from a complexity point of view, 'nested' (and operative) within it. However (and this is what makes scientific knowledge our best kind of artistry yet), what science, as a way of knowing, has absolutely built into it, *as a central principle*, is the forward-directed openness which actually marks all human endeavour, but which is also constrained by the power of tradition and myth. What this means is that science, even as its great methodological power pushes it forward into more and better knowledge about reality in a relatively unconstrained fashion, is at the same time still to some extent restrained by the belief systems within which it operates. But eventually (although 'eventually' may be centuries) its 'paradigm constraints' will be shifted – not only by the weight of accumulated evidence (as Kuhn says), but by the fact that it has forward-directed openness as a central cognitive principle of its labour on the world. In this way, albeit *relatively* gradually, science as a way of knowing is more able to accommodate cultural change than is mythic or traditional knowledge (and in many ways is itself an instigator of cultural change).

We can see an illustration of this in the relatively swift shifts in argumentational premises that Rick Rylance details in his study of these movements in the second half of the nineteenth century: 'What appeared coherent to, say, a theologically inclined thinker in 1850, appeared gratuitous and incoherent to an agnostic or "practical atheist" in 1875, because the central term holding together the description of the operations of the physical world – that is, God Himself – was no longer taken for granted by the latter'.[12] Here, change in cultural tone is marked by the passage of a mere twenty-five years. The tendency to greater openness to cultural change is marked also, perhaps, by the confidence with which Candace Pert, one of the scientists discussed later in the book, has been able to assert her belief that both the direction of her research, and its blocking over many years by a male-dominated scientific establishment, was closely tied to the fact of her gender.[16] The force of her scientific work, however, meant that eventually (in this case sooner rather than later) it was recognised; and, given its content, both scientific and cultural paradigms subsequently shifted forward. The implication is not, of course, that female scientists do their science differently, but, rather, that their research questions may be framed in different ways. And of course, and as Pert also says, the power of cultural norms is so great that 'boundary-breaking ideas are rarely welcomed at first, no matter who proposes them'.[17]

This book, then, is a sketch, by way of exemplary instances, of the history of the paradigm shift which has been taking place in the sciences for the past sixty to seventy years. Nobel Prize winning physicist Robert B. Laughlin has described this shift as the move from the Age of Reductionism to the Age of Emergence.[18] Chapter 1 outlines the long history and prehistory of the complexity revolution, tracing it back at least as far as the Romantic philosophy of the late eighteenth and early nineteenth century. Chapters 2 and 3 chart some of the history of challenges to the dualism of the Cartesian paradigm. Chapter 4 brings the account up to date with a discussion of these developments as they currently stand. The concluding chapter goes on to discuss human creativity in culture and some of the implications of understanding this in terms of complex systems processes – not least in terms of education and a non-cynical conception of what we owe to the future. I will return to a slightly more detailed description of the contents of these chapters below.

The reader will, perhaps, be unsurprised to discover that a complexity view of nature and culture suggests that a narrowly economic and utilitarian focus on markets and job-skills (and the consequent degrading of non-instrumental life in all its forms) as the main means by which the good society can be achieved is profoundly and dangerously mistaken. Throughout the book, my central concern is to show that systems, or complexity, theory represents a new scientific paradigm shift across the disciplines, which allows us to think about human societies in a different way. This does not do away with the methodological power of reductionism in science, but it provides it with, and sets it within, a larger frame of understanding. Most importantly, in showing that societies, and then cultures, are a part of biological evolution (as indeed Marx himself believed), and that our sociality is fundamental to our physical and mental well-being, this book's intention is to affirm, against the individualism asserted by liberal doctrine, the crucial importance of human solidarity and sociality.

Of course, readers wedded to the idea that everything is made in discursive language may find themselves suspicious of my commitment to the idea that there definitely is a material reality about which much is now known and more can be known – although I hope that what I have to say about the centrality of semiotics, understood more widely, will allow them to see that postmodern (largely humanities) preoccupations with the sign and signification have been *more or less* on the right track (albeit sometimes paradigmatically deformed by the ideological sea in which they were born). Equally, readers belonging to the sceptical materialist school of hard-headed facts may well find themselves disturbed by the mind-body-environment holism which I discuss here, in which emotional life is absolutely central – a fundamental insight of the contemporary scientific research I describe. People wholly given over to the belief that reason can *only* operate in a world of measurable things denuded of the affective body – a world in which such 'non-things' as subjectivity and the emotions have for so long not been granted any reality because not measurable – will find the challenges to this old way of thinking substantial and deeply unsettling. Still, this is the new science now. In the relatively new fields of neuroscience and psychoneuroimmunology, methodologically well-grounded experiments and research, repeatedly confirmed, show the physical means by which mind – i.e. environmental experience, relationships, words and images – effects emotional and physiological changes. Mind, body and environment are a processual continuum; what you feel, your emotional life and responses to where you are and how other people treat you, get quite definitely written in your body in terms of flourishing or disease.

EVOLUTION AND EMERGENCE

Some readers may feel anxious about, or openly hostile to, the idea that biology can tell us anything about culture. Such apparent 'biological determinism', they will rightly feel, simply reduces human culture to biology, and leaves no room for the considerable effects of culture on human behaviour. But this view depends, precisely, upon the idea that nature and culture are quite different things, and that a linear description of biological evolution as simple cause and effect (rather than complex cause and effect producing stratification and emergence) is a satisfactory one. This is understandable, given that until quite recently the best known attempts to take the sciences, especially biology, and in particular human biology, into the field of human motivations and behaviour have remained caught within the limitations of the reductive paradigm. Neo-Darwinists (now usually called Evolutionary Psychologists) attempt to explain all the complex evolutionary structures which evolutionary time has imposed, beyond and above the level of the gene, by reducing them to genes alone. This over-simplification has brought into disrepute more complex understandings of evolution. Though evolutionary psychologists can point to *some* (and probably only a few) genetic configurations capable of fairly direct effects – they can be definite predictors of disease, for example – most are not. Furthermore, even in cases where genetic predisposition can be reckoned significant (in schizophrenia, for example), material environmental experience (including emotional experience, as I argue in Chapter 4) remains highly significant. Evolution – physical and cultural – is a complexly (i.e. non-linear) accumulative process, and its results cannot be satisfactorily explained by reduction to earlier stages. It seems to me that a fully materialist account of human life and, importantly, human sociality will remain seriously incomplete until we are able to offer a compelling evolutionary biological account of them which is non-reductive, and has persuasive explanatory power with regard to human beings as language users, and beings who live in a complex world of signs. Sociobiology, and all forms of genetic reductionism, are, unsurprisingly, silent on this score. Yet what is *Homo sapiens*, if not the creature that lives and creates *in signs*?

When my students tell me, as they still often do, that 'reality is constructed in language', I respond by saying that it is certainly true that language is very important and powerful. It is, I say, referring them to Jared Diamond's *The Rise and Fall of the Third Chimpanzee*, responsible for nearly all the advantage we have over our nearest primate relatives.[19] Our ability to put our experiences into words, to develop higher-order abstractions from these, to analogise and metaphoricise, and to bring different abstractions together in unusual ways, is the source of our phenomenal creativity. Language is also causally efficacious; it can change the way we think and feel about things dramatically. However, I add, there are very good reasons for believing that language, for all its formative powers, cannot be the whole story about how the world is. For example, I say, how do you think you got here (and I don't mean on the number 43 bus)? After a puzzled pause, some-one usually says 'Do you mean where do humans come from?', and I say, yes, I'm asking you about human origins. Though there are exceptions to this, the usual consensus which emerges is something along the lines of 'by evolution'. And so, I push on, what does this tell us? What is the story, widely believed by most educated people and certainly by nearly all scientists, that the theory of evolution tells?

The answers which follow are, of course, that evolutionary theory tells us that humans evolved from ape-like forebears in a very long process of random (which we now know to be genetic) mutation and adaptation to environment. And although I no longer believe that this linear story of evolutionary descent by random genetic modification and adaptation is the whole story, I am generally content to let things rest at this point with the addition of one further question: what, I ask, does this, especially the 'environment' part of it, tell us about reality, or the way the world really is? The answer, of course (and my students usually get to it quite quickly), is that if environments aren't real things, and organisms aren't real things responding to real environments, evolution simply cannot occur. And so, I say, there must be a reality, in which all species evolved, which is not simply made up out of language in the sense we ordinarily mean by the word. Whatever our super-subtle minds and language have achieved, there is a substrate, of the body and its sensory ways of apprehending the world, upon which the development of verbal language and its abstractions are entirely dependent.

Of course, the question of what constitutes an environment, and how we can most usefully think about it, is not a closed one. An ecosystem, for example, is usually defined as 'the currents of energy and matter that bind individual populations within a network together'.[20] But, clearly, not all organisms have access to the environment in the same way, since the latter is a creaturely world (*Umwelt*) called forth by a sensorium, and the messages the organism can 'read' or 'understand'. And the human sensorium, although tremendously 'message', or better, 'semiosis', rich, is still limited to its own *Umwelt*, which understands things semiotically, but *not only* in articulate language. Even so, this 'reading' remains a 'heavily processed' affair; what we sense (see, hear, smell, feel, and so on) is the result of enormously complex biosemiotic processes. Biologists Humbert Maturana and Francisco Varela have compared such processes to 'one voice added to many others in a lively family discussion where reaching agreement on, for example, action to be taken will not depend on what *one* particular member of the family has to say'.[21] So, although the world is real, I am not suggesting that it is 'out there' in some simple kind of way. It is called forth in the organism/*Umwelt* continuum. Similarly, our models and descriptions (including scientific ones of course) of the world are metaphoric, and the metaphor which seems right for understanding something at one point may actually prove later to have been not quite right, or actually constraining – as will be seen to be the case later with the metaphor of the immune system as a discreet kind of border police. For more discussion on changing ideas about the immune system, and what it tells us, alongside this idea of semiosis, and molecular biologist Jesper Hoffmeyer's idea of 'semiotic freedom', see Chapter 4.

The puzzling story (briefly referred to at the beginning of this chapter) of how many intelligent people in the humanities and social sciences came to ignore the theory of evolution, and to believe that everything we think we know is just an effect of spoken or written language, is yet fully to be told – although John Deely's work, mentioned above, takes it a considerable way.[22] Its roots probably lie, at least in part, in the now infamous mind-body dualism philosophically developed by René Descartes, the further roots of which probably lie in the historical convergence of neo-Aristotelianism with Platonic idealism and Christian idealism in the Middle Ages. The result of this mind-body dualism was, paradoxically, *both* the development of a powerful objectivism, in which the world could be seized as an object of knowledge, and, also, eventually, a kind of romantic subjectivism in which 'my' knowledge seemed absolutely particular to me, and uncommon. This latter idea, of course, was fed by the individualistic world view we inherited from the development of liberal philosophy in the seventeenth and eighteenth centuries. But to say that physics and biology *also* matter is not at all to deny the great power of human verbal language. In our own time, the positive critique of phenomenology which led us to consider the abstracted 'constructed' nature of verbal (including, of course, written) language's account of reality was not a rejection of phenomenology; it was an on-going and positive critique of the 'Western metaphysics' which, for example, Derrida discovered still lurking in Husserl's phenomenology. Postmodern linguistic constructivism (and value relativism) seems to have been driven by a curious mix: on the one hand there was a desire to challenge the positivistic and instrumentalist world-view associated with modern capitalism and science; on the other hand there was the pursuit of an alternative idea of community-in-language, which was finally self-defeating because it could not make any positive claims about reality, or the desirability of any real political and social change.

Cartesian mind-body dualism and the Romantic tendency (although not in all its manifestations) towards forms of subjectivism are each problematic (human experience is powerfully collective – both 'downwards' in terms of the biology of the tribe, and 'upwards' via the development of language, culture and society): my argument is that we now stand on the cusp of finding new non-dualistic ways of talking about ourselves, through the development of the idea of complexity theory and the science which it describes. As Robert Laughlin, mentioned earlier, writes:

> Much as I dislike the idea of ages, I think a good case can be made that science has now moved from an Age of Reductionism to an Age of Emergence, a time when the search for ultimate causes of things shifts from the behaviour of parts to the behaviour of the collective. It is difficult to identify a specific moment when this transition occurred because it was gradual and somewhat obscured by the persistence of myths, but there can be no doubt that the dominant paradigm is now organizational.[23]

Here again, my reader might well like to consider the prescience of Raymond Williams's attempts to 'think the future' in terms of 'the whole process of our living organisation' in the passage from *The Long Revolution* which I quoted at the beginning of this introduction.

Williams's insights, drawing on scientific developments, indicate, I hope, the kind of 'forward-directed' nature of human knowing in culture which I outline in Chapter 2's discussion of the structure of tacit knowledge in complex systems as described by Michael Polanyi. Importantly, I also hope that these things to which I am drawing the reader's attention will go some way to giving the lie to the pessimistic idea that there is no such thing as progress. Our progress, as Hegel (the first philosopher of 'evolutionary' consciousness as systemic) describes it in *The Phenomenology of Spirit*, certainly proceeds by indirection;[24] but, complex and indirect as it is, it *is* real. A dispirited culture which has lost sight of this, which buries itself in the conviction of meaninglessness, is simply one which has given itself over, mistakenly, to the inevitability of hopelessness and cynicism. This is a stage, or stratum, in the evolution of human culture; but it is certainly not *all* that there is yet to be known and said. Only a remarkably excessive pride could see it as such.

As I have outlined above, this book is an attempt to sketch out this transition from the analytic atomism of reductionism to the synthetic and collective holism of emergence as it appears at key moments in the history of ideas. But its essential purpose is two-fold. First, I wish to do what I can to make an intervention in thinking in those parts of the humanities, cultural studies, sociology and politics which are still in thrall to postmodernism. In all four areas, the idea that what is true is *simply* what is 'made' in language is pernicious and has gone deep. In politics it has resulted in the 'lawless world' described by Philippe Sands in his book of the same name, in which facts themselves, and the fact of internationally agreed laws, for example, are concocted or ignored in preference for ideology.[25] In the humanities, cultural studies, and in parts of sociology, it has resulted in a state of affairs which Joseph Carroll, seeking to draw his own discipline of literary studies into the shared world of modern science, compares to the level of technological development in a third world country.[26] And although Carroll's prescriptions and uses of Darwinism seem quite wrong to me (because still caught within a reductionist paradigm, and concomitantly unproductive), his work is part of a move within humanities, cultural studies, and sociology scholars to rejoin their disciplines with the world of science. The impulse to bring together again the 'two cultures' is now very widespread.

But second, and more importantly, I wish to make a good materialist and – as far as my abilities allow – scientific case for regarding the philosophy and ideology of neo-liberal individualism, as understood and promoted from the 1970s onwards, as profoundly and damagingly mistaken. This is not to suggest that the greater individualism, and sense of the importance of the individual that has in some ways marked the modern epoch is some kind of a mistake which can be repaired by reverting to an ill-considered collectivism. Neither is it to suggest that individualism is a new, specifically Western European, accomplishment of the last two and a half centuries. It isn't.[27] Rather, I wish to take issue with two oddly related ideas which came to dominance in the 1970s: the neo-liberal idea that 'there is no such thing as society, only individuals and their families' (as infamously stated by British Prime Minister Margaret Thatcher); and the 'postmodern' idea that 'reality is constructed in language'. The first idea was tied to a politics which wished to promote the idea of human beings as primarily selfishly motivated economic actors in a world conceived of nearly entirely in terms of the marketplace. The second idea, originally believed to be emancipatory, counted only verbal language as a worthy object of communication studies thus effectively cutting off human culture from nature and from the sciences which try to describe it.[28]

The connection between these two enterprises can be found in their inclination to ignore – amongst a host of other real things – the biological reality of human sociality. In the first case, the impetus was clearly to subordinate all political and social arguments to the contrary to the argument for the overwhelming dominance of market relations above all other relations; and, in the second case, the driving force (apart from academic career-building – which is equally responsible for some nonsense in the sciences, of course[29]) was an intellectual argument, that all discourses are constructed 'power/knowledges' (Foucault), which do not describe reality (which has no independent existence outside of language), and can, thus, be endlessly deconstructed (Derrida) to reveal their ideological nature.

Both these views of the world were radical, and regarded as progressive by their proponents on the political right and left respectively. Both are essentially wrong, or at least seriously incomplete. The 'marketisation' view of human beings has been massively socially disruptive and damaging in its subordination of all human needs to those of the market.[30] The worst effects of the 'discourse' view have lain in the ways in which such value relativism has insinuated itself into the culture in the form of the belief that all that matters is what is written or said ('represented' – as in political 'spin' for example), and that language has no firm relation to facts in an actually existing real world.

The etymology of 'radical', of course, leads back to the Latin *radix*, meaning 'root'. Sometimes 'being radical' can mean going down to the root of something in order to pull up the whole tree because it has grown deformed, and it is necessary, thus, to clear the ground for newer and healthier plantings. On other occasions, though, the radical gesture may lie in the ecological discovery that the root system supports many other forms of life in the forest, and that to tear it up without thought for its wider consequences has incalculably worse long-term effects. For me, J.S. Mill's twin essays on Bentham and on Coleridge remain an exemplary statement of this modern dilemma.[31] The purpose of the present book, then, is to argue for the reality of the world, especially the reality of human biology and its consequences, and to seek a fuller understanding of articulate language as the most recently evolved expression of a semiosis which informs *all* nature. In this, the semiotic nature of human beings will prove central – both in drawing attention to the fundamental nature of their sociality, and in showing its evolutionary and ontological dependence upon natural and cultural environments.

We all know that what it is possible to think is limited by the ideas and models available. What is important about complex systems theory and science is that it provides us with a new way of thinking, not only about how complex systems work, but about how, in their biological manifestations and beyond, they are *inter-related*. Particularly with the relatively recent development of biosemiotics in the work of Thomas A. Sebeok and, more recently, Jesper Hoffmeyer, we can begin to see that complex systems theory actually tells us something important about life itself.[32] A full description of a complex biological system is, at the same time, a description both of all forms of organism, from the most complex down to the single cell at least, *and also* a description of human language – more properly understood as semiosis. As Sebeok has said, the definition of life may coincide with the definition of semiosis. Once we recognise that all complex systems, in biology *and* in culture and society, are characterised by the self-similarity (iteration) of nature's patterned parsimony, it makes perfect sense both that an organism's *Umwelt* (its particular environment) is its space of semiotic experience (semiotic niche), limited only by its species-needs, and that the evolution of life can be understood as the story of the emergence of increasing levels of ever more semiotic and *Umwelt* complexity – or as Hoffmeyer puts it 'semiotic freedom'. Importantly, understanding that semiosis is not peculiar to humans, but 'goes all the way down', so to speak, allows us to think about the range of verbal and non-verbal semiosis that constitutes human experience. We do not, thereby, in any way lose the ability to consider the power of abstract conceptual language; rather we add to this a greater understanding of our own verbal and non verbal (averbal) communications and tacit knowledges as these inform our relations to others, to our own *Umwelts*, and to the social and natural environments we share.

The current time, of Laughlin's Age of Emergence, is a particularly propitious time for such a re-engagement between 'the two cultures' because what complexity science, and its identification of phase-shifts and emergent levels, shows us quite clearly are the ways in which evolution, from the basic laws of physics all the way up through chemistry, biology and the emergence of language, culture and society in *Homo sapiens*, is a continually evolving story of the fundamental *physical* relatedness of space, matter and life. Even the vexed question in physics about the relationship between the quantum universe of instability and the stable world described by Newtonian physics can now be understood, via elegant and reliable experiments described by Laughlin, to be explained in terms of 'the emergence of Newtonian reality out of quantum mechanics'.[33]

Darwin's theory of evolution (although it is incomplete, as is suggested both by recent genetic research and biosemiotics, in which – in both fields – organisms are no longer understood as passive subjects of random mutation but as active construers of their environments[34]) takes us a considerable way in our understanding of biological evolution and complexity. But, of course, once we enter the realm of human self-consciousness, culture and society, it inevitably becomes much more difficult both to track the story of emergence and also to agree about what does and doesn't constitute such epochal change. This is because phase-shifts, and emergent phenomena, like semiotic phenomena, are always context-dependent, that is to say 'environmental' processes. The context in which, to cite one of the experiments described by Laughlin, sound waves revert to quanta is that of ultralow temperatures. The chemical phase-shift between water and ice is similarly temperature context dependent. Where human beings are concerned, we are all aware of such emergent phenomena as the ways in which, for example, the life of the group, tribe, society is not reducible simply to the sum of its individual parts; but identifying the moment that a collection of individuals becomes a group is likely, again, to depend upon the contexts in which they find themselves. Passengers on a boat, or aeroplane, for example, may only very weakly, if at all, experiences themselves as a gestalt; but, should their vehicle be hi-jacked, they are much more likely to experience the emergence of a collective identity in which the group has a life and emergent intelligence of its own.

These questions about complex social totalities are interesting and possibly hard to answer. Certainly they open up a host of further question; but, before giving up on them as incapable of resolution, it may be helpful to consider the history of social totality, or systems, thinking as that developed in the context of nineteenth- and twentieth-century science and of scientifically informed thought. In the course of doing this, it should become possible to consider the ways in which science, and particularly the 'softer' end of it – biology, psychology, sociology, anthropology and so on – has been distorted by an exclusive focus on reductionism and the mechanistic metaphor introduced by seventeenth-century European science. It is noticeable that (though there have, of course, been considerable advances, for example in molecular biology) it is the life and social sciences that have suffered most from this kind of approach. Reductionism has expanded enormously our knowledge of biology, and, as in the science of genetic modification, our ability to interfere with biological life, but it is still the case that absolutely no-one understands with any precision how the complex life of the cell uses DNA and mRNA to 'tell' some cells to turn into, say, the liver, and others to turn into the heart or the skin. As biologist Stuart Kauffman says:

> Despite ... the brilliance manifest in the past three decades of molecular biology, the core of life itself remains shrouded from view. We know chunks of molecular machinery, metabolic pathways, means of membrane biosynthesis – we know many of the parts and many of the processes. But what makes a cell alive is still not clear to us. The centre remains a mystery.[35]

In the arts and humanities, of course, a sense of the emergence of real creativity (i.e. true novelty), from both the patient rehearsal of skills over time and from a particular receptive attentiveness, has preserved these kinds of disciplines and practices from both reductionism and mechanism. I discuss the continued importance of the arts and humanities (including contemporary religious interests and contemporary literary/cultural critical developments in eco-criticism and semiotics) in Chapter 3.

COMPLEXITY AND HUMAN SOCIALITY

The whole argument of this book, then, depends upon an understanding both of complexity theory as currently conceived, and also of its proto-history in human ideas. Marx and Engels were the first thinkers to draw upon romantic philosophy in order to attempt a critique of the bourgeois philosophy of possessive individualism, and of the liberal philosophy and political economy which it produced, and Chapter 1 commences with a brief discussion of their work. The reason for this opening emphasis is my desire, also indicated above, to offer an intervention in contemporary political (in the broad sense) theoretical debates. Part of the current problem in political thought stems from the hybridisation of liberalism and socialism in much modern thinking. And although socialism is essentially about the social and collective nature of human beings, the inexorable rise in individualism (which is perhaps, deep down, as much about our yearning for our lost souls as anything else[36]), and the liberal idea of personal freedom (and the codifying of rights and obligations), have increasingly tended to occlude the equally important fact that human beings are, indeed, social, i.e. collective, creatures.

The effect of liberal ideology, and its general occlusion of the power of human sociality, has had the effect of placing responsibility for human flourishing (and, in its absence, blame) upon individuals, when, in fact, it is quite clear that flourishing (and its absence) is, at base, systemic, not individual. The liberal argument to the contrary is pervasive and ideological; it is driven by the interests of particular dominant groups (we used to use the term 'class', but this has proven too simple), and is also the source of a great deal of human suffering. A large part of what I hope to achieve with this book is to enlist the new science of the 'Age of Emergence' in order to demonstrate how, from a newly developing scientific point of view, the philosophy, science and economics of liberalism are both progressive (we pay greater attention to individual cases and to the rights of individuals) and also, ultimately, not quite right (we under-estimate the force of collective life). As Robert Laughlin says:

> Ironically, the very success of reductionism has helped pave the way for its eclipse. Over time, careful quantitative study of microscopic parts has revealed that at the primitive level at least, collective principles of organisation are not just a quaint side show but *everything* – the true source of physical law, including perhaps the most fundamental laws we

know. The precision of our measurements enables us to confidently declare the search for a single ultimate truth to have ended – but at the same time to have failed, since nature is now revealed to be an enormous tower of truths, each descending from its parent, and then transcending that parent, as the scale of measurement increases. Like Columbus or Marco Polo, we set out to explore a new country but instead discovered a new world.[37]

And just as reductionism has paved the way for a new consideration of complex collectivities in the sciences, so, similarly, social and political atomism must stand in need of a similar reconsideration of human collectivities and sociality.

Since the 1970s, we have come to understand more and more about the importance of complex systems and about how they work over both long and short periods of time. As I have intimated, much of our understanding about how complex systems develop over long periods of time has come from evolutionary biology. Biology also has taught us important lessons about the nature of the intimately 'coupled' relationship between organisms and their environment. This includes humans, for whom environment is both natural and cultural. The co-evolution of organisms and environments involves, of course, very long periods of time. On much shorter time-scales, from the spread of viruses, to the spread of computer 'viruses', to the spread of cultural ideas, innovations, fashions and fads, and so on, the idea of webs of connections (uneven, often 'hierarchical', and focused around what Albert László Barabási calls 'hubs'[38]), characterised by the features of complex systems everywhere, such as 'phase-shifts' and emergence, remains central. These sorts of relatively short-term activities of complex systems are now widely studied across a range of disciplines, from management theory to marketing, and political science to epidemiology. In fact, epidemiology has some very interesting things to teach us, not only about the importance of natural and social environments, and our fundamentally social and collective nature, but also about the power, health, and life and death expectations which attach to social hubs.

As well as these short and long term timescales, I also deal with ideas about complex systems over the medium term, which is much shorter than biological evolution (but towards which the latter obviously contributes), but much longer than fads or fashions or social trends. Sometimes we use the word 'civilisations' to describe these medium term developments, but these can also be thought of collections of epochs, or of generations. In fact all the words we use to describe periods of historical time can be thought of loosely as our attempts to describe the ways in which social and cultural forms emerge out of each other: (some) fads into fashions; (some) fashions into trends; (some) trends into generational phenomena; (some) generational phenomena into the flavour of eras and epochs, and so on. Over time there is a constant process of web-like but uneven interconnection and emergence which produces stratification. We recognise stratification when we note historical change in all its many manifestations. Our difficulty lies in describing it, most especially so because it is this mix of society, culture and environment (both human and 'natural').

Part of the problem of description lies in our modern habits of reduction and atomistic analysis. These were powerful conceptual tools in the development of Newtonian physics of cause and effect, but they have not served us well in our understanding of complex systems, where causes are far too complex to model with linear mathematics, and can only be understood by noticing the behaviour of a system *as a whole*. In social systems, for example, individual illness cannot simply be understood individually. As distinguished physician and epidemiologist Michael Marmot puts it, 'There is almost no condition in medicine that has a single cause';[39] in fact, it turns out that the factor which overwhelmingly, and in every study, predicts your health and life expectancy is the richness of your place in the web of connections, your status, and thus your power to influence events around you. The evidence is overwhelming. Add to this what psychoneuroimmunology has discovered, and we can begin to put together the 'mechanisms' joining together inner and outer worlds; but these are in fact very far from mechanistic, as was not so long ago (about twenty years) imagined. Unsurprisingly (remember Nature's patterned parsimony, or what Charles Sanders Peirce called 'nature's tendency to take habits'[40]), 'inside' resembles 'outside', and just as nature, and spectacularly human social life, outside us is full of semiosis, so is inside. What were, not so long ago, believed to be the three separate systems of the body (nervous, endocrine, and immune) turn out to be one integrated complex system of messages. We are a system of systems, or 'swarm intelligence' as Jesper Hoffmeyer puts it.[41]

The observation of stratification is particularly clear in scientific and technological development. What I am particularly interested in, though, are the forces (creativity might be a word that describes part of this) which drive emergence and stratification. These forces are partly individual (inasmuch as they often emerge from the work of talented individuals) but they are overwhelmingly social and collective – hence the widely observed phenomenon of 'simultaneous discovery'.[42] Creativity, the production of innovation as opposed to simple newness, whether in the arts, humanities or sciences, is the result of our collective and co-operative natures. When scientists talk about objectivity, they mean at least two things: the attempt to eschew, as far as possible, prejudiced or 'taken for granted' ways of thinking, but also the *collective* confirmation of data by other people with the training and equipment to do so.

Eschewing 'taken for granted' ways of thinking is actually quite hard. Poets and other artists know this as well as do scientists, because human thought is heavily dependent upon language (which is why scientists like mathematical ways of describing things; the precision aids mental discipline). In many ways human creativity *is* innovation in language. Ann Senghas's study of the evolution of Nicaraguan Sign Language confirms this,[43] as do anthropological studies such as those by Merlin Donald and Luigi Luca Cavalli-Sforza.[44] The long hegemony of liberal individualism, supported by the economic and imperial power of the West, can be understood as the hegemony of a certain language (based on a philosophy) which is the taken for granted common sense of that hegemony. In Chapter 2, I draw on Michael Polanyi's argument about the nature of personal knowledge – as impassioned and characterised by the embodiment of tacit knowledge that cannot be described or transmitted in the abstractions of spoken and written language – in order to argue the case for the *political* need to re-evaluate what currently counts as knowledge, objectivity, and 'fact'.

Indeed, it is the entire argument of this book that liberalism's over-emphasis upon individualism, and upon a certain idea of human reason and knowledge, is mistaken. Not only is our understanding of human behaviour and reasoning incomplete when we fail to take account of the role of the body, emotion and tacit knowledge, but it is also incomplete, as suggested above, when we view people primarily simply as individuals. It is the *whole* creature (mind-body-environment) and the *whole* system (minds-bodies-cultural-social-and-natural-environments) which must be taken into account by anyone interested in human flourishing and creative living.

In Chapter 1 I trace briefly the development of proto-complexity thought in the nineteenth century, from its roots in eighteenth-century German Romantic Philosophy (and Marx's use of these), through Darwinian theory and the development of ecological understanding, and then on through the development (again from biology) of complex systems theory in the 1930s, general systems theory and cybernetics in the 1940s and 1950s, and complexity science from the 1970s. I also offer an outline of complexity theory as it is now broadly understood.

In Chapter 2 I look at the legacy of nineteenth-century positivism, and the mind-body-dualism-induced over-emphasis on objectivity in science (and, indeed, all disciplines which seek scientific status), via complex-systems-informed-critiques. In particular I draw on the work of Michael Polanyi – with his insistence on the existence of tacit knowledge and the impassioned nature of the human will to know more about the world – and on Brian Goodwin's discussion of organism, and human, flourishing in terms of the idea of 'generative fields' (*Umwelts*).

In Chapter 3, I turn to a discussion of human intimations of what constitutes a good, i.e. flourishing, life, and argue that science, religion and art (broadly conceived) are not fundamentally different activities, inasmuch as each is arrived at as a way of affirming the reality of our communal experience as environmentally and socially embedded creatures.

In Chapter 4, I arrive at what is really the heart of the book, when I endeavour to show how we can find a common *biological language* (of the biological semiosis found in epidemiological research into the *social* sources of biological illness and mortality rates, for example, and in the recent development of the discipline of psychoneuroimmunology) that can also offer an account of the evolution of *cultural language*. This takes me into a biosemiotic account of complex systems and emergence, as suggested by Charles S. Peirce's semiotics; and by Jakob von Uexküll's biology (although von Uexküll seems to have had no knowledge of this work). It also takes me to Thomas A. Sebeok's twentieth century developments in semiotics in the second half of the twentieth century; and, finally, to the further elaboration of biosemiotics and complex evolutionary systems found in the work of molecular biologist Jesper Hoffmeyer in the late twentieth and early twenty-first centuries. With this, the understanding of culture as the evolution of nature – not as something fundamentally different – is achieved. In an echo of Hoffmeyer, I do not wish to return human culture to nature (the reductionist neo-Darwinist move), but to return nature to human culture.

I say that Chapter 4 is the heart of the book because it is here that I make my final arguments for the biological, and psycho-biological, basis of human sociality. Properly understood, this claim about complex biological systems and their intrinsic relation to complex social systems and relations, has wide-reaching political implications. What we learn from complexity theory and science is that human creatures simply cannot be properly understood as the isolated, rationally choosing, self-maximisers so beloved of liberal politics and political economy. We learn that 'mind' cannot be understood simply as mental events going on inside individual heads; it is, powerfully and *really*, in our bodies, in the world, and in other people. Subjectivity is intersubjectivity.

In my concluding Chapter 5, I argue that a politics of full human cultural flourishing must be based on the maximisation of general creativity (and such cultural flourishing includes the natural flourishing of the biosphere, as will be shown in Chapter 4, through an account of the essential organism-environment coupling expressed in von Uexküll's conception of the *Umwelt*). I offer a speculative, but hopefully well-informed, account of the human evolution of semiosis *as culture*, drawing largely, but not exclusively, upon the work of Hoffmeyer. And, returning to the concerns of Chapter 3, I emphasise the importance of the evolution of forms of creativity in representational techniques and their accompanying conceptual architecture. I refer to Margaret Boden's extensive research on creativity and artificial intelligence in order to emphasise, if such should be needed, that creativity is a *social* phenomenon which depends upon access to social connectedness and consequent influence. This chapter includes a brief discussion of the AI development of neural nets and parallel-processing, and the ways in which these developments are helpful to an understanding of Hoffmeyer's use of the notion of 'swarm intelligence', and of human beings, and societies, as 'swarms of swarms' (and 'swarms of swarms of swarms') in the form of nested complex systems. In chapter 5 I also suggest (drawing on Lyn Margulis and Hoffmeyer, and perhaps research to be elaborated in further work) that the fundamental biological relation, and its negotiation, is that between self and other, or identity and difference. I suggest, thus, that this is the basic creative gesture in the evolution of all life – and that, in the evolution of human society and culture, this takes the fundamental form of the ethical relation.

At present, collective creativity certainly exists, but it is only generally recognised as such, and politically affirmed, in the form of corporate creativity exercised via markets in which it is commodified. Following Karl Polanyi, I argue that, while all known societies have used markets, no society before Europe in the eighteenth century has made them central to human affairs.[45] Indeed, like Marmot, Karl Polanyi argues that status – i.e. the *collective* recognition of your *social* significance – is the main motivator in human behaviour. I argue, thus, that there are not only good socialist reasons for seeing capitalism, and the individualistic philosophy which supports it, as a hugely successful ideological feint that empowers the social significance of the few against the needful claims of the many; but there are also good biological and scientific reasons for seeing things this way.

NOTES

1. R. Williams, *The Long Revolution*, Peterborough, Ontario: Broadview Press Encore Editions, unrevised reprint of London: Chatto & Windus, 1961, pp13-14.
2. L. Margulis, *The Symbiotic Planet: A New Look at Evolution*, London: Phoenix, 1999.
3. Williams, *The Long Revolution*, ibid., pp36-7.
4. E. Jablonka and M.J. Lamb, *Evolution in Four Dimensions: Genetic, Epigenetic, Behavioral, and Symbolic Variation in the History of Life*, Cambridge MA: MIT, 2005.
5. Jablonka and Lamb, *Evolution in Four Dimensions*, ibid., p1.
6. See, for example, L.A. Pray, 'Epigenetics: Genome, Meet Your Environment', *The Scientist*, Vol. 18, Issue 13/14, July 5 2004; also at https://notes.utk.edu/Bio/greenberg.nsf/0/b360905554fdb7d985256ec5006a7755?OpenDocument.
7. Williams, *The Long Revolution*, ibid., p9.
8. Williams, *The Long Revolution*, ibid.
9. Williams, *The Long Revolution*, ibid., pp38-9.
10. J. Deely, *The Impact on Philosophy of Semiotics*, South Bend, Indiana, St.

Augustine's Press, 2003; J. Deely, *Four Ages of Understanding: The First Postmodern Survey of Philosophy from Ancient Times to the Turn of the Twenty-First Century*, Toronto: Toronto University Press, 2001.
11. J. Deely, *The Impact on Philosophy of Semiotics*, ibid., p.89. See also, J. Deely, 'The Impact of Semiotics on Philosophy', www.helsinki.fi/science/commens/papers/greenbook.pdf.
12. K. Soper, 'Postmodernism, Subjectivity and the Question of Value', J. Squires, ed., *Principled Positions: Postmodernism and the Rediscovery of Value*, London: Lawrence & Wishart, 1993.
13. R. Williams, *Culture and Society: 1780-1950*, London: Hogarth Press, 1987 [Chatto & Windus, 1958].
14. T.A. Kuhn, *The Structure of Scientific Revolutions*, London: University Of Chicago Press, 1962.
15. R. Rylance, *Victorian Psychology and British Culture 1850-1880*, Oxford: OUP, 2000, p205.
16. C. Pert, *Molecules of Emotion: Why You Feel the Way You Feel*, London: Pocket Books, 1999.
17. Pert, *Molecules of Emotion*, ibid., p19.
18. R.B. Laughlin, *A Different Universe (Reinventing Physics from the Bottom Down)*, New York: Basic Books, 2005, p208.
19. J. Diamond, *The Rise and Fall of the Third Chimpanzee*, London: Vintage, 2002.
20. J. Hoffmeyer, *Signs of Meaning in the Universe*, tr. Barbara J. Haveland, Bloomington: Indiana University Press, 1996, p59.
21. H. Maturana & F. Varela, *The Tree of Knowledge*, Boston: Shambala, 1987, cited in Hoffmeyer, *Signs of Meaning*, ibid., pp71-2.
22. J. Deely, *Four Ages of Understanding*, op. cit.
23. Laughlin, *A Different Universe* op. cit., p208.
24. As J. N. Findlay puts it in his Forward to the *Phenomenology*, 'the lessons that consciousness learns in its continued experience of objects are not *for it* a continuous course of lessons: it conceives that it is constantly passing to some new and unrelated object, when it is really only seeing its previous object in some novel, critical light'. G.W.F. Hegel, *Phenomenology of Spirit*, tr. A.V. Miller, Oxford: OUP, 1977, pxv.
25. P. Sands, *Lawless World: America and the Making and Breaking of Global Rules*, London: Allen Lane, 2005.
26. J. Carroll, *Literary Darwinism: Evolution, Human Nature, and Literature*, London: Routledge, 2004, px.
27. J. Goody, *Capitalism and Modernity: The Great Debate*, London: Polity, 2004.
28. P. Cobley, ed., *The Routledge Companion to Semiotics and Linguistics*, London: Routledge, 2001.
29. Laughlin, *A Different Universe*, op. cit.
30. See, for one example among many, R. Sennett, *The Corrosion of Character: The Personal Consequences of Work in the New Capitalism*, London: W.W. Norton & Co., 1998.

31. J.S. Mill, 'Bentham' (1838) and 'Coleridge' (1840) in J.S. Mill and J. Bentham, *Utilitarianism and Other Essays*, ed. A. Ryan, Harmondsworth: Penguin, 1987.
32. T.A. Sebeok and J. Umiker-Sebeok, *Biosemiotics: the semiotic web*, The Hague, Mouton de Gruyter, 1991; J. Hoffmeyer, *Signs of Meaning*, op. cit.
33. Laughlin, *A Different Universe*, op. cit., p107.
34. E. Fox Keller, *The Century of the Gene*, London: Harvard University Press, 2000.
35. S. Kauffman, *Investigations*, Oxford: OUP, 2000, p2.
36. Jeremy Waldron argues that the foundations of John Locke's liberal egalitarianism were, indeed, deeply theological. J. Waldron, *God, Locke and Equality: Christian Foundations In Locke's Political Thought*, Cambridge: CUP, 2002.
37. Laughlin, *A Different Universe*, op. cit., p208.
38. A-L Barabási, *Linked: The New Science of Networks*, Cambridge, Mass: Perseus, 2002.
39. M. Marmot, *Status Syndrome: How Your Social Standing Directly Affects Your Health and Life Expectancy*, London: Bloomsbury, 2004, p10.
40. Hoffmeyer, *Signs of Meaning*, op. cit., p27.
41. Hoffmeyer, *Signs of Meaning*, ibid., p113*ff*.
42. M.A. Boden, *The Creative Mind: Myths and Mechanisms*, 2nd ed., London: Routledge, 2004, p45 [Wiedenfeld & Nicolson, 1990].
43. A. Senghas and M. Coppola, 'Children creating language: How Nicaraguan Sign Language acquired a spatial grammar', *Psychological Science*, 12, 4: 323-328; Senghas, R. J., A. Senghas, and J. E. Pyers, 'The emergence of Nicaraguan Sign Language: Questions of development, acquisition, and evolution' in J. Langer, S. T. Parker, & C. Milbrath (Eds.), *Biology and Knowledge revisited: From neurogenesis to psychogenesis*, Mahwah, NJ Lawrence Erlbaum Associates, 2004.
44. M. Donald, *Origins of the Modern Mind: Three Stages in the Evolution of Culture and Cognition*, London: Harvard UP, 1991; L.L. Cavalli-Sforza, *Genes, Peoples and Languages*, tr. M. Seielstad, Harmondsworth: Penguin, 2001.
45. K. Polanyi, *The Great Transformation: the political and economic origins of our time*, Boston: Beacon Press, 1957 [1944].

CHAPTER 1

A long time coming: the complexity revolution

During the twentieth century, the idea that social change is an evolving process became unfashionable in sociology. In the nineteenth century, as with the work of Marx and Engels, for example, there was a much stronger sense of society as a complex and evolutionary process, and this was linked, via the hopes and aspirations of such thinkers, to an ideal of progress. In the late nineteenth century, and in the twentieth, an idea grew up that society should be conceived of as essentially static, or as seeking always to return to a state of equilibrium which could be reductively analysed into component parts. The reasons for this are, themselves, complex (and probably derive from a complicated mixture of an idea of 'proper' positivistic science alongside a sense that 'science' is not necessarily, itself, simply 'progressive').[1] In the last two decades of the twentieth century, and under the banner of postmodernism and cultural studies, the idea of processual change (of evolution as cultural as well as natural) was equally rejected, because processual change was elided with an idea of progress felt to be scientistic, possibly ethnocentric, and certainly tainted by Enlightenment rationalism.

In fact, and as Norbert Elias's *The Civilizing Process* demonstrates at length, human societies clearly are subject to something very like cultural evolutionary processes, in which we move from lesser to greater limitations upon our instinctive desires, and from lesser to greater subtleties of understanding of ourselves. As is the case with all evolutionary developments, human societies evolve greater cultural complexities. This is 'progress' to the extent that we understand ourselves better – but this should not be confused with the idea that technological development alone means that we understand ourselves more clearly; the relationship between technological development and social and individual development is, at best, uneven. Social and individual development is, clearly, related: what a society finds needful is expressed in the modes by which each individual is socially disciplined according to predominant structures of feeling.[2] Technological development, and the ways in which it is understood, is both a product of social development and, always, an open move beyond it. Science and technology in a technological society is socially produced, but always contains also the seeds of another way of understanding things.

This is the situation in which we find ourselves in the early years of the twenty-first century. The positivistic sociology of the twentieth century was countered by both continuing marxist analyses and by structuralism. Indeed, Terry Eagleton has argued that virtually all so-called 'contemporary Theory' is, in fact, an ongoing 'friendly' critique of marxism.[3] The post-structuralist move, especially with Jacques Derrida's deconstruction, was essentially to point out that the idea of structure, itself, remained too static. Similarly, his critique of 'western metaphysics' did not involve the claim (as has often been thought) that 'everything is constructed in language', but, rather, the detailed phenomenological observation that propositional, abstract, conceptual language is always capable of formulating nonsense-knowledge – especially when it is divorced (as it tends to be in scientific modernity) from the experiential knowledge, borne of the body, but in which 'we know more than we can tell'.[4]

But while 'theory' in cultural studies and the humanities was trying to maintain the spirit of earlier forms of critique capable of grasping the social whole, science itself was undergoing an evolutionary (one might say almost revolutionary) phase-shift in the form of the development of open, evolving, non-linear systems theory (sometimes called General Systems Theory; now often simply referred to as complexity science). With this, and eventually, the whole fortress of philosophical dualism, the metaphysics of scientific objectivity and the liberal idea of the individual as isolated monad must gradually crumble. It is not, of course, that the argument (that the absolute distinction between individuals and societies is false, and that societies are processual) is new: Norbert Elias's *The Civilizing Process* makes it forcibly, as does Raymond Williams's *The Long Revolution*. However, it is the first time that science has developed, in a fullish way, an account of phenomena in terms of open, complex, evolutionary, self-organising adaptive systems. With this, our social self-understandings – indeed, our whole dominant modern way of understanding individuals and societies – is surely set to change. Certainly, the *Homo oeconomicus* of liberal political economy (along with the psychology which that proposed) that has been so dominant in Western countries in the last nearly three hundred years, will not survive unscathed. Neither, as sociology increasingly understands society as a complex evolving processual whole, will the currently dominant political idea that social relations can be replaced by quasi-market relations remain tenable.

The development of science, remarks Elias, is dependent on the structure of feeling achieved by people at any historical time. The development of modern science depended on a more detached viewpoint (and an increase in internalised control of passionate life) generally.[5] Thus, as he goes on to say, as people achieved this degree of removal of themselves from the centre of things, which allowed the conception of, for example, a heliocentric rather than geocentric universe, it is unsurprising that they were, at that time, more concerned with the quality of the object of knowledge than of the subject doing the knowing. We can, thus, see the gradual development of complexity science over the past fifty years as, itself, a reflection of a developing capacity to reflect on our embodied and socially embedded nature as selves. And this scientific development can be seen to take its place within and alongside other similar developments – in object relations theory in psychoanalysis, in structuralism in anthropology, in Elias's work in sociology (*The Civilizing Process* was originally published in 1939), in structuralism and post-structuralism in linguistics, philosophy and the humanities more generally, and in every discipline which has attempted some theorisation of subjectivity as not simply monadic, but relational and social. Complexity science is, in many ways, a corrective to the science which has hitherto held sway. Its implications, for how we think about knowledge, individuals, society, the sciences and the arts, are very vast.

It is, though, in politics that the gradual coming into popular awareness of complexity science will have its most momentous effects. Capitalism, as is well known, is very wedded to the idea of people as essentially economically striving individuals. Socialism, which opposes this view of things, is fundamentally a recognition of social complexity, without, until recently, the science to back it up. At the core of left progressive thought lies the conviction that society is not composed of atomised individuals primarily pursuing their own economic self-interest, but, rather, that societies are to be understood as complex totalities. Along with that conviction goes the related one that governments are, or should be, institutions for intervening in complex totalities in strategic ways so as to ensure the health, creativity, adaptability, and generally fullest flourishing of all the parts (humans and their systems) of that totality. No government has ever succeeded very well in that task (even where that was its professed aim) because no-one has ever formulated a sufficiently comprehensive understanding of how complex totalities actually work. Generally, and inasmuch as they have been driven by scientific theories and data rather than by conviction (good or bad), governments have thus fallen (happily or not) into the larger ideology of liberalism which philosophically informed modern science and actually constitutes European modernity itself. I shall come back to this.

In the last fifty years, though, a new kind of science has been gathering: the science of complex totalities. We are now fairly well on the way to understanding how such open, evolving, adaptive, self-organising (or autopoietic) non-linear systems work. Much work remains to be done, and Stuart Kauffman has recently suggested that complexity science has now reached its teenage years.[6] This new science confirms some core socialist, and green, insights – human societies are not mere aggregates of selfish and competing individuals but living totalities in which the life of the whole exceeds the sum of the parts; individuals are phenomenologically embodied and embedded in culture and nature which must be seen as inextricably intertwined and co-dependent and co-evolving. The Western emphasis on individualism needs to be understood within this wider context – and the new science also undermines some other cherished beliefs, such as the possibility of any kind of simple or reductive determinism. The fact that complex systems are not simply deterministic, but rather are complexly overdetermined, does not, however, mean that strategic interventions in them are not possible. Far from it. However, and as more or less everyone really knows, the relation between causes and effects in complex systems is not linear. In order to make effective interventions in complex biological systems such as ecologies and human societies, we need to have a good theory, one that is consistent with complexity science's general observations about the characteristics of complex systems, and describes the nature of agency (in humans conscious and unconscious) in all biological systems. But a further exploration of such a theoretical approach must wait until Chapter 4.

Before I go on to draw out some of the extensive work which remains to be done in elaborating more fully the ways in which complexity science aids an understanding of human culture and society, I want to go back to what can be thought of as the evolving history of complexity thinking in western philosophy and science (which can hardly be disentangled); it is in this way that we can begin to understand the evolutionary phase-shift that contemporary developments in the field represent. For complexity science brings along with it not only a different way of understanding open systems (ecologies, societies, organisations and so on); it also brings about a way of thinking, acting, and thinking about thinking and reason, that is quite different from 'western metaphysics' – i.e. the dominant philosophy and science of European modernity. It thus challenges the 'common sense' of the Enlightenment and scientific and industrial revolutions; our world is immersed in this western metaphysics so completely that it is invisible to us. Or, rather, some of its contradictions are visible to us – but only in ways which appear to us as feelings and intuitions, and which do not (or have not until recently begun to) count as scientific knowledge. Only a change from within science itself is capable of shifting that metaphysic. I will be arguing that complexity science presents the beginning of precisely such a momentous shift.

In what follows in this chapter, I restrict myself to western philosophy and science because, although complexity thinking has long existed in eastern and other philosophies, such philosophies have been generally understood by most commentators to consist of insights which are essentially of a religious type. This is also true of some of the most important and influential systems of thought in western philosophy, but this dimension of the thought has tended to be occluded in direct proportion to the secularisation of western society, and to the extent that religious thought is believed to be in conflict with materialist understandings and claims to scientific truth. I will return the question of religious, or sacral, thought in Chapter 3.

PREHISTORY OF COMPLEXITY

There is a quite long history in European modernity of variously (although not very dissimilarly) expressed anxieties about social alienation. Indeed, in essence, this is what romanticism is. Starting with Kant's attempt towards the end of the eighteenth century to describe the kinds of human thought (rational, moral and aesthetic), and the limits of these, romanticism (Hegel, the Schlegels, Schiller, and a host of other German philosophers – in Britain, Coleridge, Wordsworth, and later Keats, Shelley, Byron and so on) represents the beginnings of the phenomenological revolt against the predominance of the kind of thinking which seemed to fuel the scientific revolution of the seventeenth and eighteenth centuries.

This thinking was a mixture of Baconian empiricism and Cartesian rationalist dualism. Combined, these led to the idea of a distinct separation between, for example, mind and body, subject and object, affect and reason, and the human and natural worlds; and the effectiveness of science was increasingly believed to reside in its objectivity and in its method of reduction to parts. The world was conceived via the dominant metaphor of the machine – very often a mechanical object such as a clock – the workings of which could be comprehended by taking it to pieces and analysing the working of all the parts. That this reductive method (including, as far as possible, the isolation of the part from the potential contaminations of the whole) was so successful, and that the world as described by physics was deterministic and could be rendered mathematically in linear equations, was, of course, extremely powerful. It promised new understanding, predictive power and control. Its consequences were, though, to give excessive weight to the importance of abstract, conceptual thought, an atomistic view of things, and the kinds of knowledge these can achieve, but to do so at the expense of occluding (fatefully, phenomenology would say) the very great importance of relational understandings and experiential and tacit knowledge – in which, as Michael Polanyi put it, 'we know more than we can tell'.[7] This attempt to reduce all phenomena to *one* level of behaviour and explanation, understandable in the early days of science, has persisted with tremendous tenacity – largely, one surmises, because it supports a particular ideological view of society. And although physics itself went

1. A Long Time Coming 43

through its own proto-complexity revolution nearly a hundred years ago with the discovery of the non-linear world of particles, and the realisation of different strata, atomistic thinking still persists both in physics and in some corners of biology (population genetics, and neo-Darwinism, for instance) in the form, for example, of arguments which attempt to explain the organism by reduction to the gene.[8] In other important ways, however, biology has been at the forefront of the complexity revolution in the sciences.[9] I will address this neo-Darwinism, and also the effects of complexity science in biology, in Chapter 2.

Before the development of complexity science proper, however (and the understanding that open systems evolve, via emergence, strata of ever greater complexity over time), the first response to this scientific reductionism took the form of romanticism. What we have come to call romanticism consists essentially of all those thinkers in whom we recognise the confraternity of the thought that says that complex wholes (such as societies and their natural and historical contexts) are not to be understood by reduction to parts, that the whole is greater than the sum of its parts, and, relatedly, that real knowledge does not consist simply in objective, abstract, propositional knowledge, but is also, and just as importantly, deeply experiential.

That this is true must be admitted by anyone who is not seriously deluded. Human experience is limited to the world as afforded by the evolved human sensorium. It is certainly not the same experience or world as that afforded to other creatures with differently evolved sensoria. What the world looks like 'objectively' – i.e. from no creaturely view at all – is, and must remain, a complete mystery. Relatedly, we can know that our concepts arise from empirical experience situated in time because we know very well that the concepts (and abstract thought) that we develop from this would simply be unimaginable (or at least very hard to imagine) for people alive before these various aspects arising from our experience and science had occurred. What to people of the past is unknowable, unthinkable, mysterious or an object of religious veneration may, for us, be something common and well understood – at least in our experience and use of it, if not necessarily in our abstract conceptual knowledge of it. We also know that this will be true for people in the future who will understand things about the world which to us are mysterious or simply conceptually unavailable now.[10]

Nonetheless, and incredible as it seems to many, this vast forgetting, or disregarding, of the body, and of experiential knowledge, is the basis upon which much modern philosophy has proceeded.[11] In concentrating on the kind of abstract conceptual and propositional knowledge which also appears powerfully to guide modern science (I shall come back to that 'appears' later), and in playing, very seriously of course, with all the things that conceptual language can do when it forgets its embodied source, modern philosophy has managed to assert, as truly possible theories of how the world is, things which are ludicrous: for example, that a machine based on silicon technology and binary notation could develop a mind and intelligence similar to carbon-based biologically evolved human beings. This is the kind of thinking which Husserl called 'metaphysical'.[12] Nonetheless countless hours have been spent by perfectly intelligent people thinking in such a fashion, and millions and millions of pounds and dollars have been invested in the possible applications of such thinking – in cognitive neuropsychology and in AI, for example.

That romantic philosophers were very often interested in art, or were themselves actually poets, is to be explained by their understanding of the embodied and tacit nature of skilful creation. The semantic distinction between artisan and artist arises at just this time, and it surely does so because of the desire to make a distinction between skilled creative artisanship, which is definitely non-utilitarian, and other kinds of skilled artisanship which may be. In his *On Naïve and Sentimental Poetry* of 1795-6, in which he contrasts the 'spontaneous' poetry of the ancients with the 'reflective' poetry of the moderns, Friedrich Schiller understands the modern poet's task as precisely a conscious overcoming of the split between conceptual and experiential thought brought in by modern philosophy and science in the seventeenth century. In *On the Aesthetic Education of Man* (1801 rev'd), Schiller commented on the meagre and fractured experience of modern labour and life, in which full participation in the whole of life was increasingly circumscribed by the development of machine labour and the dictates of procedures conceived along utilitarian lines.[13] Later in the nineteenth century, the demand for non-alienated labour of all kinds would be extended by Carlyle, Ruskin and William Morris.[14]

COMPLEXITY AND SOCIALISM

The history of complex totality thinking is, of course, also a large part of the thinking that has informed socialism. It is very clearly what Engels thought he and Marx had been doing when he summarised their endeavour in his 1886 essay on Feuerbach. In that essay he makes quite clear their understanding of culture and society as an evolutionary development of natural evolution as described by Darwin. In particular, he outlines their understanding of the difference between evolution in nature and evolution in society. In the former, the forces of change are 'only blind, unconscious agencies acting upon one another, out of whose interplay the general law comes into operation'. In the case of social evolution, obviously, things are quite different: 'In the history of society, on the contrary, the actors are all endowed with consciousnesses, are men acting with deliberation or passion, working towards definite goals; nothing happens without a conscious purposes; without an intended aim'. Nonetheless, he goes on to say, whilst there is a demonstrable evolutionary progress in society and, thus, 'the course of history is governed by inner general laws' (in which men discover new things about nature – including themselves – and culture; first tools, then more advanced technologies, all of which tend to bring about change at every level – social and cultural), it is also the case that individual willing is often governed by 'accident', which 'apparently reigns on the surface': 'that which is willed happens but rarely; in the majority of instances the numerous desired ends cross and conflict with one another, or these ends themselves are from the outset incapable of realisation, or the means of attaining them are insufficient'. With the advent of class society, however, in which the economic well-being and flourishing of the bourgeoisie requires the oppressive exploitation of the working class in ways that are fundamentally opposed to their economic well-being and flourishing, the forces of history become easier to comprehend, exposed as they then become as the antagonism of class where the agency lies in class consciousness and class action.

The problem with this compelling analysis (and its being carried through) is, as I intimated in the opening paragraph of this chapter, that the philosophical, and hence scientific, air in which it was expected to breathe, live and have its being was dominated by a scientific worldview thoroughly opposed to such holistic, or complex totality, thinking. In order to 'get heard' in the nineteenth and twentieth centuries, it was necessary to put on the clothes of science and 'objective' materialism. Eighteenth-century philosophy and science was dominated by an atomistic view of the world in which society was composed of striving, economically self-interested, individuals. In this context, the essentially socialist impulse which evolved at the end of the eighteenth century, and manifested itself in large part in romantic philosophies, did so in a context in which its completely correct understanding of human society as a complex evolving whole must either find a 'scientific' (i.e. objectivist and positivistic) basis or, as in fact generally happened, retreat into the sphere of the aesthetic. The term 'romantic', which arose in the seventeenth century in association with the creative freedom of imagination found in romance writing, gradually acquired its negative connotations (except in the sphere of the affections); it is now generally used as a way of indicating that something is either unworldly or unscientific. But this just indicates the power of the invisible ideological sea in which we swim.[15] The power of Marxist philosophy was, however, too great for it to be tarnished by its origins in the thought of Hegel, and, by the end of the nineteenth century, it had called forth, in Europe but not in Britain, the response of bourgeois philosophy in the form of sociological systems theory.[16]

Socialism found its scientific clothes, but at the cost of using a language at odds with its own best insights, and a language, furthermore, which tended, over time, to corrupt its early senses and understandings. An example of scientific socialism using abstract thought at the expense of experiential thought is the view that the death of millions of people is worthwhile if it achieves a truly socialist society of perfected equal relations. A skilfully developed experiential view would counter that human societies are not the kinds of things which can be made perfect or absolutely equal. Individuals may become saint-like, but no entire human society ever will. Such an idea is abstract and disembodied in the extreme. No-one should be deemed sacrificable on the basis of it: it is inhuman.

A complex totality view of society as it has emerged in the science of the last fifty or so years would, though, bring about a number of very welcome changes in our ways of thinking about the world – some of which, in relation to the social sciences, have been discussed by David Byrne.[17] Not least, and as Byrne argues, is its emphasis on the importance of experiential knowledge, and the local and particular as opposed to the universal and abstract, a complexity-led view of things would tend towards democratic decentralisation and a valuing of local knowledges. Equally important, though, is Byrne's use of the idea of 'control parameters'.[18] Complex systems can be, non-deterministically, 'worked' (largely through interventions in the form of positive and negative feedback) in all sorts of ways; we are all familiar, for instance, with the ways in which a single strategic intervention in a meeting can sway the direction of subsequent understandings and events. What matters is the driving force behind the intervention (its political intent) *and* the understanding of interconnectedness, non-linearity, emergence and autopoiesis (self-organisation). Where, as in Byrne's example, the control parameter is 'the degree of inequality', this would mean a decisive shift in real power and resources in which both these latter were really put in the hands of disadvantaged people at the grassroots level, who would be given the freedom to self-organise local initiatives, projects, and so on. Examples of this kind of 'bottom up' empowerment are given in Brian Goodwin's discussion of the 1930s 'Peckham Experiment', and other more recent developments. Goodwin writes:

> There are signs that regeneration is actually occurring on the basis of the type of holistic, bottom-up approach that worked so well in the Peckham Experiment, such as is described in *Radical Urban Solutions* by Dick Atkinson, a central figure in the St Paul's project in Birmingham. Here, housing associations, self-help day centres and local enterprises are interacting co-operatively with church, school, and community in one, and there are plans for a neighbourhood bank. This is grass-roots politics that exposes most current political ideology as irrelevant. It re-empowers communities 'to take back what's ours by right', as one parent put it. At this point we enter politics and the necessity for a new economic order, which is beyond my scope. Hazel Henderson's *Paradigms in Progress* (1993), Helena Norberg-Hodge's *Ancient Futures*

(1992), Edward Goldsmith's *The Way* (1992) and James Robertson's *Future Wealth* (1990) are among the books that point the way to a different future in which co-operative relationships and quality of life become primary components of social and economic activity.[19]

ABSTRACT CONCEPTUAL KNOWING VERSUS EXPERIENTIAL EMBODIED KNOWING

Before moving on to a consideration of complexity science as it is currently understood, perhaps it is necessary to say a little more about conceptual knowledge (abstract intellectual knowledge *that*), experiential knowledge (phenomenological embodied knowledge *how*), and tacit knowledge, in which latter 'we know more than we can tell'. Tacit knowledge is creaturely skilful phenomenological knowledge. Human creatures *know* they have it, and exult in its expression (as skilful being in the world), and in wondrous reflection upon this knowledge – which cannot be put into words, but which is experienced in all creative artisanship and art, and in creative and skilful living generally. This is language as semiosis which is not reducible to words, but which is embodied in acts. When Raymond Williams, in his essay 'Crisis in English Studies', ventured the idea that the development of semiotics as a 'system of signs (not confined to language)'[20] might be useful, he was surely thinking of the usefulness of systems of 'reading' which do not reduce knowing to knowing in conceptual linguistic terms, but in which we can talk, for example, of 'reading' the combination of gesture, rhythm, tone and space in a dance, or of colour, brushstroke and content in a painting. He was concerned, as others have been before and after him, with those kinds of knowledges which are embodied in lived and skilful engagement with the world and with other embodied creatures. These are *not* knowledges which can be easily accommodated in a propositional language, because they are knowledges borne of the body – which is more or less dumb in terms of propositional, abstract, knowledge, but still *really* knows what it knows.

Tacit knowledge is real knowledge semiotically conveyed and 'read', but it cannot ever be wholly conveyed in propositional language; it is a skilfulness (for good or bad) which can only be learnt by practice in doing (apprenticeships of various kinds; all children learning from their care-givers). Nonetheless, it is this kind of embodied knowledge (which abstract, conceptual, knowledge cannot ever describe in all its fullness) which turns out to be central to understanding the human knowledge of complex totalities. Skilful being in complex totalities is what humans (and all creatures) have evolved to do. Skilful being in cultural complex totalities is a specifically human skilful being in the world. Actions (especially, perhaps, political actions) driven mainly by abstract thinking, which forget embodied experience, local knowledge and skilfulness, are always, almost by definition, dangerous. Most people, even when they cannot say *why* they know it, do know this to be true. In allowing us a scientific conception of complex totalities and the kinds of embodied, experiential, intersubjective, skilful knowledge by which they are negotiated, complexity science thus throws a counter-weight against the kind of abstract propositional knowledge carried out by supposedly atomistic individuals which has dominated western science and philosophy for the past three hundred years: 'three centuries of subjectivist madness', as Bertrand Russell put it.

Tacit, semiotic knowledge might begin to look like a good candidate, then, for understanding interactions in complex biological systems, provided that the argument can be made for semiosis as a feature of *all*, and not only human, biological systems, and as 'going all the way down' to the simplest elements of biological life. But, before I can come on to a fuller discussion of this it is necessary to pass through a number of other discussions on the way: a more detailed discussion of complexity science, its emergence (particularly in biology), and the characteristics of complex systems; a deeper discussion of the emergence of a critique of scientific objectivity and the relationship between tacit knowledge and scientific discovery in the work of Michael Polanyi; and a further and more detailed discussion of more recent developments in complex biology in the work of Brian Goodwin, Humberto Maturana and Francisco Varela. All this will be accompanied, pretty continuously, with the attempt to keep in mind the social, cultural and political implications of these discussions.

Of course, *all* spoken and written language (although not all knowledge) is, by necessity, conceptual and propositional; articulate language is bound to bow to the 'excess of world over word', and to develop abstractions from that profusion; a language which tried to name every single thing and instance would be unintelligible. When I claim that the problem with modern western science and philosophy as it develops from the seventeenth century onwards is that it is dominated, as never before, by a conceptual thought radically divorced from experiential thought, it might be objected that, surely, Christian and Platonic thought are similarly so dominated. I think that the straightforward response to this must be that, *in practice*, this was not the case. In Europe, it is not until the late Middle Ages (significantly from around the period of the Reformation in the 1530s) that the advent of conduct literature, which is aimed at the taming of spontaneous bodily behaviour, indicates the beginning of a widespread drive towards what will become the conduct of the proper bourgeois body.[21] The development of a concern with manners from the sixteenth century to the late eighteenth century suggests that the precondition of cartesian philosophy was that it was able to draw on an older dualistic tradition which was being reinforced by a generalised Protestant mistrust of sensual pleasure and adornment, and festivals of misrule and bodily excess, which the Roman Catholic church had been able, via the partial formalisation of popular carnivalesque impulses into Church feast days and masses, to incorporate.[22] With this, the body was not simply tamed, but was more or less expelled as a source of knowledge about the world.

ART, SCIENCE AND PASSIONATE KNOWING

Suspicion of the passionate 'animal' body has a long history in both philosophy and religion. In the renewed humanism of the European Renaissance, for example, this led to a sustained and anxious debate about the difference between humans and other animals, as the work of Erica Fudge, for example, has demonstrated.[23] But the question about what bodies – and especially human bodies (or body-minds) – can learn and know, and the structural movement of that knowing, did not fully begin to be explored until Michael Polanyi took up the problem in *The Tacit Dimension*. This, itself, was a rearticulation of the problem which Husserl had identified, in *The Crisis of European Sciences and Transcendental Phenomenology*, as modern science's evacuation from knowledge of lived experience.[24] Experiential, phenomenological, knowledge – the not wholly self-present or self-conscious knowledge of a body in the company of a self-reflexive mind capable of nurturing it – clearly is knowledge upon which we can quite consciously draw. But it hovers somewhere between the experiential cunning of the animal and the more self-disciplined and *attentive* cunning of the man. It partakes, Polanyi will say, in a structure of knowledge (arguably discoverable also in the structure of evolution itself) of 'disattending from' what is accomplished (or 'symbiotically' internalised) and 'attending to' something else: what it might next be useful to assimilate, perhaps? I will come back to this later in the chapter. At this point, anyway, we can see that biological and intellectual assimilation might be quite closely related, by the same natural logic.

Milan Kundera's account of philosophy's and science's loss of the life-world (*die Lebenswelt* – the place, so to speak, where experiential and conceptual knowledge are lived together), which Husserl's pupil Heidegger described as 'the forgetting of being', leads us to Literature, particularly in the form of the modern novel, as the place where, in modernity, experiential knowledge continues to investigate itself as inseparable from the intellect:

> The crisis which Husserl spoke of seemed to him so profound that he wondered whether Europe was still able to survive it. The roots of the crisis lay for him at the beginning of the Modern Era, in Galileo and Descartes, in the one-sided nature of the European sciences, which reduced the world to a mere object of technical and mathematical investigation and put the concrete world of life, *die Lebenswelt* as he called it, beyond their horizon.
>
> The rise of the sciences propelled man into the tunnels of the

specialised disciplines. The more he advanced in knowledge, the less clearly could he see either the world as a whole or his own self
[...]
Perhaps it is Cervantes whom the two phenomenologists neglected to take into consideration in their judgement of the Modern Era. By that I mean: If it is true that philosophy and science have forgotten about man's being, it emerges all the more plainly that with Cervantes a great European art took shape that is nothing other then the investigation of this forgotten being.[25]

But that is, of course, another story. Polanyi's aim was to pursue his sense of the crisis, and to challenge the assumptions about objectivity and mind-body dualism which had occasioned it, in terms of modern science itself.

In fact, it is not at all hard to imagine an effective empirical science developing in the absence of an untoward emphasis on objectivity. It is perfectly possible to imagine the subjectively impassioned pursuit of empirically-based science – which, in itself, may well involve creative hunches, shared with others, concerning data, its checking through appropriate methodologies and observations and so on. And, of course, all this must involve abstract conceptual reasoning alongside the skilful and tacit knowledges of scientists. The reason it is not hard to imagine this is because, as probably most scientists acknowledge, this is in fact, and creatively, how science actually gets done. As Polanyi, who was himself a scientist, notes, science, as much as any other human activity, depends upon personal and impassioned tacit knowledge. Such knowledge in science

> is not made but discovered, and as such it claims to establish contact with reality beyond the clues on which it relies. It commits us, passionately and far beyond our comprehension, to a vision of reality. Of this responsibility we cannot divest ourselves by setting up objective criteria of verifiability – or falsifiability, or testability, or what you will. For we live in it as in the garment of our own skin. Like love, to which it is akin, this commitment is a 'shirt of flame', blazing with passion and, also like love, consumed by devotion to a universal demand. Such is the true sense of objectivity in science...I called it the discovery of rationality in nature, a name which was meant to say that the kind of order which the discoverer claims to see in nature goes far beyond his understanding; so that his triumph lies precisely in his foreknowledge of a host of yet hidden implications which his discovery will reveal in later days to other eyes.[26]

The idea of objectivity, which produced the predominance of a positivistic attitude in science, and in all disciplines and practices which claimed scientific status, including political economy in theory and in practice, has done untold damage in human affairs. It has done so because it has allowed the kind of abstract nonsense-thought, uprooted from experiential knowledge and skilful being in the world (which I described above in terms of certain justifications for genocide, or the millions of dollars spent on AI research in search of human-like machine intelligence) on a vast worldwide scale. This kind of thinking, which arose in accompaniment to the coming to dominance of the bourgeoisie in western Europe, is ideological through and through. Its separation of objective (rational, conceptual and universal) from subjective (passionate, experiential and particular) knowledge, along with a narrowly conceived positivistic empiricism in which human being in the world is monadic and individualistic, gave rise to an account of subjectivity as the rational pursuit of individual economic self-interest which was purely in the service of bourgeois class interests. It created a powerful hegemony of 'common sense' which is just about as far from the real meaning of the term as it is possible to get. Its imposition and political implementation in the regions in which it arose was experienced as a kind of violence (especially a class violence, which it was[27]) – even where this occurred over a fairly long period (approximately fifty years – from the 1780s to the 1834 new Poor Law in Great Britain). Its colonial imposition, which occurred much faster, deserves to be counted as among the worst cultural violences of imperialism.

At least one of the values which we place in the arts is that they rehearse for us, and remind us of, the joy of sensuous experience and its skilful enactment. 'Newness' in art is found in those skilful enactments which allow us further access to the ways in which our tacit knowledges are always capable of further expansion as we discover new tools for knowing the world. New *tools* for thinking about and knowing the world can be provided in various ways (by contact, for example, with other cultures, and their various ways of representing things). In modernity, science is a particularly effective new tool; it is the very creative new tool in which we discover more about the world. It is, in fact, a new art of knowing; but it is one which tries to disavow its artistry (skilful embodied knowing) in order to establish itself as something qualitatively different. As Robert Laughlin notes, 'good theoretical physics is actually more like art than engineering and is similarly difficult to summon up on demand'.[28] While we can imagine an effective empirical science not tied to an idea of objectivity and positivism – what Brian Goodwin has termed 'a science of qualities'[29] – modern science, and its successes, have generally made such a project quite difficult. The development of complexity science, though, is beginning to change things quite rapidly. As Ilya Prigogine, one of complexity's early pioneers, has said, with this new kind of science we may well be on the way to achieving new levels of understanding about ourselves and the world in which 'the dichotomy of the "two cultures" could and should be removed'.[30] It is already having an impact in the social sciences,[31] and in work produced under the umbrella of Critical Realism.[32]

COMPLEXITY SCIENCE
As with most important scientific developments, complexity science arose in more than one place at around the same time. It was substantially prefigured in the work of C.S. Peirce, but probably its earliest development came, unsurprisingly, from biology and the work of Ludwig von Bertalanffy. Bertalanffy realised that reductivism and mechanisism left great gaps (not least the evolution of biological complexity and morphogenesis) in biology. He began writing on this problem in the 1920s, and in the late 1930s began to give his first lectures at the University of Chicago on General Systems Theory as a method that is valid for all sciences. In 1950, 'The theory of open systems in physics and biology' was published in the journal *Science*, and 'An outline of General Systems Theory' was published in the *British Journal for the Philosophy of Science*.[33] A little later, Ilya Prigogine in flow dynamics, and Norbert Wiener in cybernetics were reaching similar conclusions in their fields. As Joanna Macy writes: 'Such thinking was not isolated. A holistic, process-orientated approach was in the air, with the appearance in that same period (the period of Bertalanffy's early works – the 1920s) of Whitehead's philosophy of organic mechanisms and physiologist Walter Cannon's work on homeostasis – a concept fundamental to systems thought'.[34] Bertalanffy published widely in biology, child development, psychology and philosophy of science.[35] The publication in 1969 of *General System Theory: foundations, development, applications* offered a definitive humanistic statement of the theory of complex living self-generating systems which included a theory of mind as emergent from brain, body, environment. Other fields also were developing in similar ways. Certainly mathematics had encountered a form of systems theory by the end of the nineteenth century (in the form of Poincaré's 'three body problem'), and responded with the development of probability theory. At the same time, ecological developments arising out of Darwin's theory of evolution contributed to an early understanding of complex evolutionary adaptive systems. Complexity science arose, as described above, in chemistry and cybernetics, in the 1940s and 1950s, and, as a part of the development of systems and systems information theory, it was developed during the 1960s as an important aspect of early computational theory. The consequent development of powerful computers, and the subsequent development of cellular automata, considerably speeded up our understanding of the evolutionary patterns and behaviour of complex systems based on simple rules. The internet is, itself, an example of such a system. I discuss the ways in which the relatively recent understanding of neural nets and Parallel Distributed Processing can help us to understand something about human creativity in Chapter 5.

Although by no means all complexity science's developments have occurred in biology, its development is perhaps most easily apprehended as the biologisation of science. We have increasingly come to understand that all open complex adaptive evolutionary non-linear systems, whether in 'information' flows in thermodynamics, or in the chemistry of cell development, have the characteristics of biological evolution as first described by Darwin – although the latter, of course, had no sense of self-organisation (autopoiesis) or its importance in evolution; this has meant that the focus on the means of hereditary transfer elaborated in genetics has, until the recent application of complexity theory, tended to obscure the systemic evolution of life.[36] Not only does 'life', or perhaps, better, the world (and all its systems including human ones), have self-organising pattern, but so does any other system which imitates its qualities in being an open and dynamical system. Quite why this should be – why it should be, for instance, that open evolutionary adaptive non-linear systems of every sort imitate precisely the patterns – of stability followed by episodes of bifurcation – found in nature in nervous systems, blood circulation systems, river deltas, and trees – is something of a mystery. Why should it be that the fractal geometry found in patterns of evolution and descent, including of languages themselves, is also described in the fractal geometry of a tree? But it is so. In *Turbulent Mirror*, John Briggs and F. David Peat write, of the fractal geometry of complex systems,

> It is now clear that fractals embrace not only the realms of chaos and noise but a wide variety of natural forms which the geometry that has been studied for the last two and a half thousand years has been powerless to describe – forms such as coastlines, trees, mountains, galaxies, clouds, polymers, rivers, weather patterns, brains, lungs, and blood supplies. Just as physics had tried to lump together a vast range of subtle properties of nature under the general heading of 'chaos' or 'disorder', these most exquisite forms in nature, with all their rich detail, were ignored by conventional geometry. Consider the way the turbulence of wind and water gouges out and sculpts the starkly dreamy shapes of canyons, mesas, and undersea grottos. Do such places lack order? Mandelbrot avers that Euclidean geometry is 'dull'. In revenge he has shown that irregularity is exciting and that it is not just noise distorting Euclidean forms. In fact this 'noise' is the bold signature of nature's creative forces.[37]

Complex systems have several regular features. They are complex (not just complicated); they are highly sensitive to initial conditions; they are non-linear and iterative, and very small initial differences can produce vastly different large outcomes (the 'butterfly effect'); the 'life' of a complex system is not reducible to its constituent parts (the whole is greater than the sum of its parts); they are recursive (multiple feedback loops – both negative and positive: positive amplifies, negative regulates) and thus overdetermined – this feature of complex systems is called 'emergence'; complex open systems evolve and are 'dissipative' – every level of emergence is more complex (and energy full) than the one which has preceded it; they exhibit self-similarity (the forces of nature seem to arrive at similar 'solutions' to a number of different 'problems') – as is demonstrated in Benoit Mandelbrot's fractal geometry; they exist close to the edge of chaos (chaos is a part of complexity science), into which state complex systems sometimes fall; and they are autopoietic (self-organising). This capacity of evolving self-organising order out of chaos gives the title to the book Prigogine wrote with Isabelle Stengers: *Order Out of Chaos: Man's New Dialogue with Nature*, and, along with self-similarity, it is perhaps one of the most astonishing discoveries we have made about complex systems.[38]

COMPLEXITY AND POLITICS

What are the implications of complexity science for political theory? The person who has contributed most to a complexity understanding of society is philosopher of science Roy Bhaskar, who very early grasped the implications of work being done in the field in the 1970s, and attempted to formulate this work (now generally understood under the rubric of 'Critical Realism') in terms useful for the social and political sciences.[39] However, Bhaskar's work on stratification and emergence in scientific knowledge is not particularly illuminating anthropologically or semiotically in ways which might be directly graspable politically.

Clearly, the observation that societies (like languages) are complex non-linear evolving holisms, in which subjectivity is relational and embedded in both cultural and natural systems, gives the lie to both the individualist and dualist argument. And although tacit experiential knowledges in systems are local, complexity science also teaches us that, as with the hologram (and also with fractal geometry), each part contains an image of the whole – an idea which is supported, for example, by Richard Wilkinson's research on health which shows that the health of *every* individual is affected by the differences in wealth in the society as a whole: the greater the gap between rich and poor, the worse the health of each individual member.[40] Presumably, there are other parameters to which such relational conceptions can be applied; Farhad Dalal, for example, has used the insights of complexity, Winnicottian psychoanalysis and group therapy in order to rethink racism.[41] David Will uses the idea of emergence ('stratification' and 'emergence' in Bhaskar) in order to argue that the emergence of different theories in psychoanalysis (Freud, Klein and Lacan, for example) are not best thought of as quite different or rival accounts but, on the basis of emergent stratification, evolved responses in which the later theory is able to illuminate more fully the concerns and relationship of the earlier ones. Thus, the Lacanian tripartite scheme of symbolic, Imaginary and Real can be seen to correspond to Freudian preoccupations (symbolic) and Kleinian preoccupations (Imaginary).[42] Complexity observations of emergence and stratification can, similarly, allow us to see that what initially seemed to be opposed accounts of the world (atomistic reductionism and romanticism) are better understood as complexly evolved strata of explanation.

A complex totality and tacit skills understanding of human societies should also make us sharply reject any account of human behaviour which depends (as does utilitarianism, for example) on the assertion that conscious reasoning is the prime motivator of human action. As many (including Bhaskar) have argued, human rationality is poorly understood. *Reasons* are always informed by knowledges of which we are not conscious, and some of these knowledges (childhood trauma, for example) may be things we could never become conscious of – although they are undoubtedly causally efficacious. Many of our knowledges are experiential and come to us in the form of feelings which cause us to make the choices we do, and act in the ways that we do.[43] Some people refer to this kind of non-propositional knowledge as 'emotional intelligence'.[44] Skilful being in the world consists in performances and actions done on the experiential basis that they 'feel' appropriate;[45] and although they come from embodied experience, we cannot say that such skilled behaviour is *only* or *simply* embodied, because mind is an emergent feature of brain/body/environment; skilfulness is the art of body and mind; skilled thought, for example, often arises in a most opaque way in the meeting between conceptual and non-conceptual thought – it *seizes* us with the same degree of passion that Polanyi describes. However, all this should not be taken to imply that tacit knowledge or understanding is simply 'good'. One can, after all, be a skilful deceiver or thief. Tacit, embodied knowledge is the stratum (or substrate) from which articulated language and abstract knowledge emerges. The former always informs the latter; but the latter intervenes powerfully in the former also. The point is not to elevate, say, emotional intelligence over articulate intelligence, but to understand much better the nature and *effects* of their interactions.

Complexity science, thus, describes a world with which we are all actually, and in practice, quite familiar. In their tacit knowledges about how the world works, human beings already know a lot about open complex systems. That this sort of knowledge is the natural knowledge borne of our creatureliness, and is, in reality, a part of our common sense (in the real meaning of the term) as a species, might suggest that, combined with our vast capacities for abstract thought, it is a right kind of knowledge. Unless we are entirely given over (and some people are, of course) to the ideology of 'western metaphysics' in which all knowledge is objective, propositional knowledge of a world which exists 'out there' independently of our sensorium, and in which subjective experiential knowledge doesn't count as knowledge at all, we will quite readily recognise the world of complexity as our own. What is different about complexity is, of course, that it arises from science itself, and that it is able to describe, mathematically and in terms which modern science understands, what had heretofore not been describable in such a form. This introduces a new way of thinking scientifically about our being in the world, and also, perhaps, a new methodological emphasis on communication.

As I hope to show in the chapters which follow, an understanding of, first, the ways in which human knowing is (like all biological evolution) structured via stratification and emergence, and, second, the ways in which (like all biological evolution) it is processual, embodied and enworlded in semiotic communication, can lead us towards an understanding of complex systems, and human life within them, which does offer a different way of framing our thinking in ways which might be directly graspable in political (and policy) terms. But, in order to get to that place (described with reference to empirical research and biosemiotic theory in Chapter 4), it is necessary, first, to turn to an account of the nature of human knowing in complex systems. Michael Polanyi describes this in terms of the human awareness of 'rationality in nature'. I will describe processual being, via a discussion of mythic knowledge and Eastern philosophy, in Chapter 3. But first, a discussion of Polanyi's idea of rationality in nature and the structure of tacit knowledge is necessary.

NOTES

1. See, N. Elias, Appendix 1: 'Introduction to the 1968 edition' in N. Elias, *The Civilizing Process: The History of Manners and State Formation and Civilisation*, tr. E. Jephcott, Oxford: Blackwell, 1994, p181*ff*.
2. Elias, *The Civilizing Process*, ibid.
3. T. Eagleon, *After Theory*, London: Allen Lane, 2003, p34.
4. M. Polanyi, *The Tacit Dimension*, Routledge & Kegan Paul, London, 1967.
5. Elias, 'Introduction to the 1968 Edition', *The Civilizing Process*, op. cit.
6. S. Kauffman, Foreword to A. Keskinen, M. Aaltonen & E. Mitleton-Kelly, *Organisational Complexity*, FFRC Publications 6/2003, www.tukk.fi/toto/Julkaisut/pdf/Tutu_6_03.pdf
7. Polanyi, *The Tacit Dimension*, op. cit.
8. For a fuller discussion, see 'Introduction' to M. Midgley, *Science and Poetry*, London: Routledge, 2001.
9. See, for example, B. Goodwin, *How The Leopard Changed its Spots: The Evolution of Complexity*, London: Weidenfeld & Nicolson, 1994.
10. B. Magee & M. Milligan, *Sight Unseen*, Oxford: OUP, 1995.
11. Perhaps we can see, in this, the body's association (for men) with women and the prototypical Eve. Certainly, seventeenth-century natural philosophers considered their new science to be masculine, and Joseph Glanville,

for one, made the connection quite explicitly. See Midgley, *Science and Poetry*, op. cit., p50.
12. Jacques Derrida, whose thesis was on Husserl's phenomenology, claims that Husserl's philosophy retains traces of metaphysical thought. But, interestingly, and whether because of his own refusals to be pinned down into a system, or out of ironic wilfulness, many of Derrida's followers have managed to make of deconstruction another bourgeois metaphysics in which the body is just a 'text' among other texts. For a helpful essay on Derrida's relationship to phenomenology, see D. Moran, 'Jacques Derrida: From Phenomenology to Deconstruction' in Moran, *Introduction to Phenomenology*, London: Routledge, 2000.
13. For relevant extracts from both these works, see D. Simpson, *On the Origins of Modern Critical Thought: German aesthetic and literary criticism from Lessing to Hegel*, Cambridge: Cambridge University Press, 1988.
14. This history of English romanticism is, of course, admirably detailed in Raymond Willliams's *Culture and Society: 1780-1950*, op. cit.
15. Where complex totality thinking takes a hold politically in a context which remains dominated by abstract thought and narrow empiricism, the outcome is likely to be totalitarianism. The German Romantic philosopher Fichte (author of *The Closed Nation State*) was enormously influential during the nineteenth century and, thus, upon those men and women who formulated National Socialism in Germany in the 1930s.
16. P. Anderson, 'Components of the National Culture', in A. Cockburn & R. Blackburn (eds.), *Student Power: Problems, Diagnosis, Action*, Harmondsworth: Penguin (in association with *New Left Review*), 1969. Orig. published *in New Left Review* in 1968.
17. D. Byrne, *Complexity Theory and the Social Sciences*, London: Routledge, 1998; D. Byrne, 'The Politics of Complexity: Acting Locally Matters', *Soundings* 14, Spring 2000.
18. Byrne, *Complexity Theory and the Social Sciences*, ibid., p146*ff*.
19. Goodwin, *How The Leopard Changed its Spots*, op. cit., pp203-4.
20. R. Williams, 'Crisis in English Studies', in *The Raymond Williams Reader*, ed., J. Higgins, Oxford: Blackwell, 2001, p262.
21. See Elias, *The Civilising Process*, op. cit.; and M. Bakhtin, 'Introduction' to *Rabelais and his World*, tr. H. Iswolsky, Cambridge Mass: MIT Press, 1984.
22. As the historian Eamon Duffy argues, in tearing apart religious institutions and relations, the English Reformation tore apart social institutions and relations also. E. Duffy, *The Stripping of the Altars: Traditional Religion in England 1400-1580*, New Haven, Conn: Yale U.P, 1992.
23. E. Fudge, *Brutal Reasoning: Animals, Rationality and Humanity in Early Modern England*, Ithaca: Cornell University Press, 2006, forthcoming.
24. E. Husserl, *The Crisis of European Sciences and Transcendental Phenomenology*, tr. D. Carr, Evanston, Illinois: Northwestern University Press, 1970.

25. M. Kundera, 'The Depreciated Legacy of Cervantes', *The Art of the Novel*, London: Faber and Faber, 1988, pp3-5.
26. M. Polanyi, *Personal Knowledge: Towards a Post-Critical Philosophy*, London: Routledge & Kegan Paul, [1958] 1962, p64.
27. See K. Polanyi, *The Great Transformation: the political and economic origins of our time*, op. cit., for a discussion of this as regards the Speenhamland Laws, the 1832 great Reform Act and the 1834 new Poor Law in Britain.
28. Laughlin, *A Different Universe*, op. cit., p87.
29. Goodwin, *How The Leopard Changed its Spots*, op. cit.
30. I. Prigogine, http://www.nobel.se/chemistry/laureates/1977/prigogine-autobio.html
31. D. Byrne, *Complexity and the Social Sciences*, op. cit.
32. www.raggedclaws.com/criticalrealism/
33. L. von Bertalanffy, 'The theory of open systems in physics and biology', *Science*, 111: 23-29, 1950; 'An outline of General Systems Theory', *British Journal for the Philosophy of Science*, 1 139-164, 1950.
34. J. Macy, *Mutual Causality in Buddhism and General Systems Theory: the Dharma of Natural Systems*, Albany: SUNY Press, 1991, p72.
35. Bertalanffy's bibliography is available on the website of the International Society for the Systems Sciences at www.isss.org/lumLVB.htm.
36. See, for example, Keller, *The Century of the Gene*, op. cit. Population geneticists such as Richard Dawkins in *The Selfish Gene*, Oxford: OUP, 1976 cannot offer any account of morphogenesis or the complex development of life on this planet since, for them, creatures are just vehicles for gene replication in which the organism-environment relation is conceived in the fairly simple way of conventional Darwinism. Dawkins's popularising of 'the selfish gene' hypothesis has been seen by many commentators as ideologically of a piece with the neo-liberal ideological temper, including sociobiology, of the period (the early to mid 1970s) during which it was conceived.
37. J. Briggs & F. David Peat, *Turbulent Mirror*, London: Harper & Row, 1989, pp90-91.
38. I. Prigogine & I. Stengers, *Order Out of Chaos: Man's New Dialogue with Nature*, London: Flamingo, 1985.
39. See Andrew Collier, *Critical Realism: An Introduction to Roy Bhaskar's Philosophy*, London: Verso, London, 1994 for an overview.
40. R. Wilkinson, *Unhealthy Societies: The Afflictions of Inequality*, London: Routledge, 1996.
41. F. Dalal, *Race, Colour and the Processes of Racialization: New Perspectives from Group Analysis, Psychoanalysis and Sociology*, London: Brunner-Routledge, 2002.
42. A summary of David Will's work can be found in Collier, *Critical Realism*, op. cit., p217ff.
43. A. Damasio, *Descartes' Error: Emotion, Reason and the Human Brain*, London: Picador, 1995.

44. D. Goleman, *Emotional Intelligence: Why it can Matter More than IQ*, London: Bloomsbury, 1996.
45. For a description of the acquisition of this kind of skilful being in relation to learning to play jazz piano, see David Sudnow's *The Ways of the Hand: A Rewritten Account*, Foreword by Hubert Dreyfus, Cambridge, Mass: MIT Press, 2001.

CHAPTER 2

'That eye-on-the-object look': complex culture and the passionate structure of tacit knowledge

> If the time should ever come when what is now called Science, thus familiarised to men, shall be ready to put on, as it were, a form of flesh and blood, the Poet will lend his divine spirit to aid the transfiguration, and will welcome the Being thus produced, as a genuine and dear inmate of the household of man.
> William Wordsworth, Preface (1802), *The Lyrical Ballads*

I want to turn, now, to a more detailed discussion of Michael Polanyi's account of tacit knowledge. What the idea of tacit knowledge demonstrates is the stratified and emergent nature of forms of human knowledge. From this it should be possible to put the phenomenological 'flesh and blood' onto, or into, complexity science. The chapter will be completed by a discussion of complexity and biology, drawing on the work of Brian Goodwin.

At the beginning of his short book (actually originally a collection of lectures) *The Tacit Dimension*, Michael Polanyi describes the genesis of the research which led him towards the idea of tacit knowledge. It sprang from a conversation with Bukharin (then a leading theoretician of the Communist Party in the USSR) in 1935. Polanyi had asked Bukharin about the pursuit of pure science in Soviet Russia, and Bukharin had replied that the idea of pure science was a morbid symptom of a class society. Under socialism, the idea of pure science would disappear because a socialist scientist would spontaneously find his or her mind turning to the practical problems of the current Five Year Plan.[1] Polanyi was struck by this because it seemed to him that this denial of free philosophical thought in pursuit of scientific understanding was a denial of precisely the kind of thinking that had formulated the marxism from which Bukharin's own argument was initially derived. What could account for such an odd self-undoing?

Thinking further, Polanyi realised that Bukharin's response, and this kind of thinking more generally, was strongly dominated by two

2. 'That Eye-on-the-Object Look' 61

impulses: one was a powerful moral motive; the other was an extreme critical lucidity born of sophisticated conceptual thought:

> It seemed to me then that our whole civilization was pervaded by the dissonance of an extreme critical lucidity and an intense moral conscience, and that this combination had generated both our tight-lipped modern revolutions and the tormented self-doubt of modern man outside revolutionary movements. So I resolved to inquire into the roots of this condition.
>
> My search has led me to a novel idea of human knowledge from which a harmonious view of thought and existence, rooted in the universe, seems to emerge.
>
> I shall reconsider human knowledge by starting from the fact that *we can know more than we can tell*.[2]

The dissonance which Polanyi identified is that produced by a failure to pay attention to the ways in which our conceptual knowledge is the product not of a disembodied mind, but, on the contrary (and like the technology which it produces), the linguistic extended toolkit produced by an embodied and enworlded, indwelling, mind. Our language which – as I said in the previous chapter – is necessarily abstract and general, is, nonetheless, a tool which we have evolved. When conceptual thought cuts itself loose from its home in the body and the world, it is capable, very often, of the sort of dissonance Polanyi found in Bukharin's communist theory of scientific endeavour.

The Tacit Dimension is the book in which Polanyi sets out, most clearly, his argument for human knowledge as the progressive acquisition (through embodied practices, including, of course, conceptual ones) of tacit skills and knowledges. But importantly, in the discussion of the nature of tacit knowledge, Polanyi's philosophy points out something about the nature of reality too. Reality is an ordered ontological entity; it is what he calls 'rational'. Were this not the case, no-one could become skilled (have internalised knowledge). (You cannot become skilled in true chaos, because in randomness there are no regularities or laws to internalise: chaos in complex systems, on the other hand, is not true chaos, but is more like a revolution.[3]) But, further, the nature and structure of tacit knowledge (actually the basis for *all* our knowledge of what is true) points to the stratified nature of reality as an evolving open system in which emergence means the emergence of new strata at ever greater levels of complexity – with each strata having the double aspect which Polanyi finds in the internalisation of tacit knowledge. But before discussing this any further, I need, first, to describe Polanyi's account of tacit knowledge in a little more detail.

In the opening pages of *The Tacit Dimension*, Polanyi tells us that he is going to borrow his idea of the structures of tacit knowledge from the philosopher Austin Farrar:

> In his book on freedom of the will, Austin Farrar has spoken at one point of *disattending from certain things* for attending to others. I shall adopt a variant of this usage by saying that in an act of tacit knowing we *attend from* something for attending *to* something else; namely *from* the first term *to* the second term of the tacit relation. In many ways the first term of this relation will prove to be nearer to us, the second further away from us. Using the language of anatomy, we may call the first term *proximal*, and the second term *distal*. It is the proximal term, then, of which we have a knowledge that we may not be able to tell.
>
> In the case of a human physiognomy, I would now say that we rely on our awareness of its features for attending to the characteristic appearance of a face. We are attending *from* the features *to* the face, and thus may be unable to specify the features. And I would say, likewise, that we are relying on our awareness of a combination of muscular acts for attending to the performance of a skill. We are attending *from* these elementary movements *to* the achievement of their joint purpose, and hence are usually unable to specify these elementary acts. We may call this the *functional structure* of tacit knowledge.[4]

It should be clear that what Polanyi is referring to is something like the gestalt of an entity. When we say of a person that they 'can't see the wood for the trees', we are referring to their inability to grasp the whole (the gestalt) as a result of an over attention to the parts. An over attention to the parts is precisely what some-one who is learning, but is not yet skilled, is obliged to do. But this over attention is precisely what prevents them, until they are getting the hang of things (or as David Sudnow, in *Ways of the Hand*, describes it 'going for the jazz'[5]), from *grasping* the whole in a skilled way. I use the metaphor 'grasping' the whole', but I do not mean the kind of mastery which conquers, but rather the mastery that indwells, which is intimate with and deeply knowledgeable about that which it knows. This is the indwelling of the conscious animal which cannot say all it knows but certainly knows what it knows and that it knows. We might also notice that an over attention to the parts, and a focus on the divorced conceptual idea of knowledge as 'extreme critical lucidity', which prevents a seeing of the whole is a characteristic of modern science *in its learning phase* – what Robert Laughlin calls the Age of Reductionism. Polanyi's account of knowledge as tacit, and of the nature of reality which this implies, suggests the eventual development of truly skilled science which understands that reductionism is a necessary first getting to grips, but also that, once these parts of learning are internalised and habitual, it becomes possible to move from consideration of the parts to consideration of 'wholes', or, as Laughlin puts it, 'collectivities'. And of course, what I am arguing here is that the beginnings of this truly skilled science come with the development of complexity science which *sees* the whole, and, thus, has the potential skill to *move* in the whole.

TACIT KNOWLEDGE AS SEMIOTIC KNOWLEDGE
But I run ahead of myself a little. To the *functional structure* of tacit knowing (internalizing knowledge of the parts so that we can 'get' the whole), Polanyi also adds the *phenomenal structure* of tacit knowing. In other words, our knowledge of the whole allows us to know *more* about it. Once I am skilful in regard to certain kinds of entities, I can be skilful in all sorts of original ways. I can learn and extend my skill. When my skills are wide-ranging, I can bring together diverse skills (or entities) in wholly original ways which create completely new things. In other words, the functional structure and phenomenal structure of tacit knowing together produce the third aspect or structure of tacit knowing which is it is meaningful. This, Polanyi says, is 'the *semantic* aspect of tacit knowing'.[6] Rather than 'semantic', and for reasons alluded to in Chapter 1 which I will pursue in more detail in Chapter 4, I would prefer to use the term 'semiotic'. To use Gregory Bateson's formulation of that which is meaningful, cited by Jesper Hoffmeyer in *Signs of Meaning in the Universe*,[7] 'a sign is a difference which makes a difference': meaning (significance) springs from semiosis, and both words have the same roots in the Greek word for significance. Our knowing, as it becomes skilful and tacit, introduces us, like streams running into a river, and a river running into its delta, to a sea of real skilful knowledge about the world. This, of course, is the nature of all creativity, which springs from the bodily roots of our knowledge to 'man's highest creative powers'.[8]

When we make a thing function as the proximal term of tacit knowing, we incorporate it in our body – or extend our body to include it – so that we come to dwell in it.[9] And, of course, this indwelling in fact describes all our skilful being in the world. At the end of the nineteenth century in Germany, both Dilthey and Lipps believed that this was the mode in which we fully experienced art – as a form of indwelling in which, thus, we allow ourselves access to the author's or painter's own mind. But for Polanyi, this is true not only of the arts and humanities, it is true of our being, including our scientific being, in the world in general.[10] It is not by looking at things that we understand them, but by dwelling in them. This is manifestly true of any field, whether intellectual (e.g. mathematics), sensuous (e.g. cooking), or a mixture of both (e.g. mothering). Indeed, as Polanyi goes on to note, the idea of extreme lucidity and objectivity may even be counterproductive:

> We can now see how an unbridled lucidity can destroy our understanding of complex matters. Scrutinize closely the particulars of a

comprehensive entity and their meaning is effaced, our conception of the entity is destroyed. Such cases are well known. Repeat a word several times, attending carefully to the motion of your tongue and lips, and to the sound you make, and soon the word will sound hollow and eventually lose its meaning. By concentrating attention on his fingers, a pianist can temporarily paralyse his movement.[11]

Learned skills, or habits, are everything when it comes to internalising knowledge. Indeed, it may not be inappropriate to describe nature itself as built upon habituation – as the semiotician C.S. Peirce did. As Hoffmeyer puts, describing the evolution of first chemical and then biological organistic *umwelts*, or 'couplings':

> Physico-chemical habits became biological habits. Primitive cells were organized into endosymbiotic patterns which we call eukaryotic cells. Eukaryotic cells acquired the habit of working together as multicellular organisms which in the course of time adapted to the prevailing logic of the ecosystems. The stabilization of living conditions under this form of logic made both longevity and intelligence an advantage and, hence, the logic of the ecosystems was eventually shattered by the appearance on the scene of humanity with its formidable talent for bossing pre-human life around. But even human beings could not shrug off this knack of forming habits. Each civilization is a manifestation of the way in which a new master plan is accepted, a plan that will significantly boost or diminish the unpredictability of human thought or deed.[12]

OBJECTIVITY AND SCIENCE

> We are approaching here a crucial question. The declared aim of modern science is to establish a strictly detached, objective knowledge. Any falling short of this ideal is accepted only as a temporary imperfection, which we must aim at eliminating. But suppose that tacit thought forms an indispensable part of all knowledge, then the ideal of eliminating all personal elements would, in effect, aim at the destruction of all knowledge. The ideal of exact science would turn out to be fundamentally misleading and possibly a source of devastating fallacies.[13]

By reference to the argument of Plato's paradox in the *Meno*, Polanyi argues that if all aspects of knowledge were explicit, no scientist would ever do what scientists regularly do, which is to intuit that there is a problem in a formulation or explanation and say what it is in order to develop a research programme. How can the scientist know that there is a problem when he or she doesn't know what it is? This is the paradox that Plato notices 'He says that to search for the solution of a problem is an absurdity; for either you know what you're looking for, and then there is no problem; or you do not know what you're looking for, and then you cannot expect to find anything'.[14] Of course, the reason that people (scientists or philosophers or whatever) do, routinely, 'see' that there 'is' a problem with something, but still not quite know what it is, is, Polanyi says, is because, when we are skilled in our particular world, we have a tacit knowledge of that world; we 'intuit' the problems to be sorted out. Too much rationality (or, more accurately, the fantasy of rationality) would be a real problem here. (Plato's answer, by the way, was that we know there's a problem because we've remembered the knowledge from a previous life.) This also accounts, I think, for the fairly well-known fact that people working in the same field very often come up with similar arguments or solutions at around the same time. Where the field is held in common, the real problems (and solutions) are too:

> The kind of tacit knowledge that solves the paradox of the *Meno* consists in the intimation of something hidden, which we may yet discover. There exists another important manifestation of these mental powers. We are often told that great scientific discoveries are marked by their fruitfulness; and this is true. But how can we recognise truth by its fruitfulness? Can we recognise that a statement is true by appreciating the wealth of its yet undiscovered consequences? This would of course be nonsensical, if we had to know explicitly what was yet undiscovered. But it makes sense if we admit that we can have a tacit foreknowledge of yet undiscovered things.
>
> [...]
>
> It appears, then, that to know that a statement is true is to know more than we can tell and that hence, when a discovery solves a problem, it is itself fraught with further intimations of an indeterminate range, and that furthermore, when we accept the discovery as true, we commit ourselves to a belief in all these as yet undisclosed, perhaps as yet unthinkable, consequences.
>
> Since we have no explicit knowledge of these unknown things, there can also be no explicit justification of a scientific truth. But as we can know a problem, and feel sure that it is pointing to something hidden behind it, we can be aware also of the hidden implications of a scientific discovery, and feel confident that they will prove right. We feel sure of this, because in contemplating the discovery we are looking at it not only in itself but, more significantly, as a clue to a reality of which it is a manifestation. The pursuit of discovery is conducted from the start in these terms; all the time we are guided by sensing the presence of a hidden reality toward which our clues are pointing; and the discovery which terminates and satisfies this pursuit is still sustained by the same vision. It claims to have made contact with reality: a reality which, being real, may yet reveal itself to future eyes in an indefinite range of unexpected manifestations.
>
> [...]

> Such indeterminate commitments are necessarily involved in any act of knowing based on indwelling. For such an act relies on interiorizing particulars to which we are not attending and which, therefore, we may not be able to specify, and relies further on our attending from these unspecifiable particulars to a comprehensive entity connecting them in a way we cannot define.[15]

The *functional structure* of tacit knowledge gives rise to both its phenomenological and semantic/semiotic forms. We internalise a skill in relation to our being in the world. The skill always *involves* the world in which we indwell (our own bodies, the forces of physics, tools including the tools of language which are the tools for thinking). The reason that we can do this is that the world is what Polanyi calls 'rational'. In other words, it has an order which we intuit and can become skilful in knowing. If the world wasn't this kind of place, we could never become skilful.

Polanyi notices (as does Bhaskar following him) that the world, and life, evolves at greater levels of complexity with each evolutionary stratum. In humans, this evolution can be seen in the evolutionary stages which the embryo passes through. In the world, the development is from physics and chemistry to biology (plant then animal). This development is repeated in modern science from the seventeenth to the nineteenth centuries and then repeated at a deeper level of strata from the end of the nineteenth and during the twentieth centuries in our knowledge of quantum physics, atomic chemistry, and cellular biology. In general, and as Polanyi notes, the things we know through our tacit powers of knowing and our skilful indwelling include ...

> problems and hunches, physiognomies and skills, the use of tools, probes, and denotative language, and my list extended all the way to include the primitive knowledge of external objects perceived by our senses. Indeed, the structure of perception throws light on all the rest. Because our body is involved in the perception of objects, it participates thereby in our knowing of all other things outside. Moreover, we keep expanding our body into the world, by assimilating to it sets of particulars which we integrate into reasonable entities. Thus do we form, intellectually and practically, an interpreted universe populated by entities, the particulars of which we have interiorized for the sake of comprehending their meaning in the shape of coherent entities.[16]

Everything about our experience as human creatures tells us that this is so about the world. We become skilful as animals, and then, via the evolution (gestural, prosodic and finally linguistic[17]) of language and mathematics, we learn that our skilful being continues at a conceptual level to be the sort of thing which is capable successfully of exploring deeper and deeper, and further and further, into the nature of reality.

2. 'That Eye-on-the-Object Look'

We gradually apprehend the different strata of things, first experientially, and then conceptually, and then we understand the place of tacit, experiential knowledge and its relation to conceptual knowledge. This marks the most skilful individual human knowing, and also, slowly but evidently, the knowing of the species.

From these facts that we can surmise some other things. We can see – as did Polanyi in *Personal Knowledge: Towards a Post-Critical Philosophy* – that *all* knowledge is personal knowledge. From our own knowledge about our own acquisition of tacit skills, we must know something important about how knowledge comes to us. Not only is it, as many educationalists have argued, *participative*,[18] i.e. the passionate doing of it can be emulated but not described in conceptual terms, but it comes to us in ways that are not deliberate or deliberable. People cannot be *given* knowledge (this is the meaning of initiation; it is not knowledge which is kept secret or hidden; it is experiential knowledge, and it doesn't matter how many times it is told to the uninitiated, they won't get it until they've got it – and this means also having the sense-based equipment necessary for getting it).[19] People can learn by way of apprenticeship, but the much more appropriate metaphor here is, tellingly, organic. People are the kinds of creatures who are inclined to grow and know; it is a part of their passionate engagement with the world that they are, given the right soil and conditions, inclined to grow skilful at being in the world (for further discussion of the conditions of human flourishing, see Chapter 4). Thus, when we talk about the importance of human flourishing, we are talking, always and at all times, about the kinds of conditions in which the roots of knowledge can deepen and the creature's good being in the world can grow and extend to the fullest of its capacities.

The organic metaphor should not lead us in the mistaken Spencerian direction of the 'survival of the fittest' or 'the selfish gene' nonsense-metaphor. The origins of life itself were probably a symbiotic co-operation of two different organisms,[20] and, as biologist Brian Goodwin argues (see below), co-operation is quite as common in nature as competition. Our knowledge of the interdependency and inter-relatedness of complex systems, and of their evolution as emergent and stratified (such that each emergent stratum contains, and is ontologically dependent upon, the stratum from which it emerged) should assure us that physicist David Bohm's idea of 'the implicate order', in which, as in a hologram, each part contains the information of the whole, is near the truth.[21]

Complex systems evolve via the emergence of strata of increasing complexity. Biological evolution proceeds in this fashion, as, we have now seen, does human culture and human knowledge. Human discovery and invention – human creativity – proceeds via tacit knowledge and our sense that we are in contact with a complex reality of which there is more to be known, and in which 'what is comprehended has the same structure as the act that comprehends it'.[22] One can say that Polanyi's idea of the structure of tacit knowledge as being repeated in what is known about reality (i.e. what is grasped as a whole can be internalised and then disattended to so as to allow for a further grasping of a greater whole) is processual and dialectical. Both in the sciences and in the arts and humanities, we know that there is more to be known because we have an intimation that there is a problem; we don't know what the problem is, but we have a strong sense that there is a problem which has not yet been satisfactorily answered. And the sense that there is a problem is always accompanied by a degree of passion which animates the explorer and drives him or her on. The marxist sense of this idea of a problem is, of course, described as a dialectical contradiction.

As I argued in Chapter 1, the Enlightenment scientific idea of true knowledge as only derivable from absolute objectivity and empirical induction has been tremendously powerful. It has gathered men to it as a cold and dispassionate thing, appropriate in its method and coolly 'rational' attitude of mind to the impersonal physical forces which Newtonian physics uncovered. The intuitive and personally impassioned aspect of scientific discovery – which proceeds via much experience and tacit knowledge in a disciplinary field, and involves what we describe as the imaginative leap from disattended-to particulars to a new level at which the whole is more deeply grasped – has been downplayed. Instead, what is *scientific*, we have been led to understand, is a reduction to parts (if at all possible in a laboratory and uncontaminated by systemic openness) and a simple functionalist determinacy – which could also provide a model for the scientific understanding of open complex overdetermined systems such as societies.

We can see that complexity science finds all open systems to contain regular features – to have order or 'rationality' as Polanyi puts it, and all human open systems to be characterised by ways of knowing (tacit knowing and skill) which both demonstrate an orderliness in the world (i.e. you can't be skilful in true chaos) *and* are inclined definitely to posit that orderliness as a possibility of further and deepening knowledge of the world. We know, from our experience of internalizing a multitude of particulars in order to become skilful with the stuff of the world, and from extending our tools for knowing further and further into the world, that the stuff of the world has an orderliness to it – David Bohm's 'implicate order' perhaps. This leads to the further surmise that, considered as a complex communicative totality, humanity shares its participative knowing – first in smallish groups, and then, to the extent that communication is increased, to an ever wider extent. This, I have suggested, accounts for the well-known phenomenon of similar scientific discoveries arising at similar times from separate individuals or research teams. Darwin's and Wallace's separate arrivals at a theory of evolution offers a well-known example, but there are many others.[23] The scientific community, in which knowledge is widely shared via overlapping disciplinary competencies, offers a paradigm of human knowledge more generally. Knowledge, and the identification of 'problems', seems to unfold according to a specific orderliness.

From this, we might further surmise that any problem which presents itself as such to a large number of people is a *real* problem. The new idea of science which I have been trying to describe above would recognise this as a scientific problem, but one to be solved not by the avowed methods of the first scientific revolution, but by the methods of the second (complexity and emergence) revolution. These latter are, actually, the disavowed (in the Freudian sense) methods of the first scientific revolution, and that they were (and still are in many quarters) disavowed also tells us something about the ideological sea of western metaphysics in which modern science was born. As I suggested in Chapter 1, it is quite possible to imagine a successful science in which the passion of discovery is publicly allowable. One can imagine a world in which science is treated as just an incredibly productive kind of art, or artisan skilfulness and creativity, and in which a very sharp distinction between art and science never arises. In such a world, science would never have come to be seen as 'bloodless' and divorced from the passionate flesh of the world.

That it did, contingently, do so in our world is a result, Norbert Elias has argued, of various economic and political pressures from the thirteenth century to the eighteenth, which fashioned the modern European state and its subjects, in which the passions had to be brought under control, and greater objectivity achieved. For Elias, this internalising of self-control (first by the court, and later emulated by the bourgeoisie), and the greater objectivity it allows, is a precondition of the modern scientific mind. Perhaps we can say that the move from a passionate immediacy of indwelling to the concentration upon discursive non-violent and conceptual manipulation of power amongst a group (at first at court, then later in civil society more broadly), is another example of the structure of tacit knowledge. This gain in objectivity makes scientific objectivity possible; but a further three centuries must pass before the obvious problems of a physics and chemistry model of science when applied to the life and human sciences gradually forces a cross-disciplinary revolution, in which there is a scientific shift of focus from artificially isolated particulars to the behaviour of complex wholes. With this shift, and over a period of about fifty years, it becomes increasingly well-understood that complex systems are everywhere around us, and that our own natural and cultural evolution can be better grasped at this new level.

PARTICIPATIVE KNOWLEDGE AND A SCIENCE OF QUALITIES

> Darwinism sees the living process in terms that emphasize competition, inheritance, selfishness, and survival as the driving forces of evolution. These are certainly aspects of the remarkable drama that includes our own history as a species. But it is a very incomplete and limited story, both scientifically and metaphorically, based upon an inadequate view of organisms; and it invites us to act in a limited way as an evolved species in relation to our environment, which includes other cultures and species. These limitations have contributed to some of the difficulties we now face, such as the crises of environmental deterioration, pollution, decreasing standards of health and quality of life, and loss of communal values. But Darwinism short-changes us as regards our biological natures. We are every bit as co-operative as we are competitive; as altruistic as we are selfish; as creative and playful as we are destructive and repetitive. And we are biologically grounded in relationships which operate at all the different levels of our beings as the basis of our natures as agents of creative evolutionary emergence, a property we share with all other species. These are not romantic yearnings and utopian ideals. They arise from a rethinking of our biological natures that is emerging from the sciences of complexity and is leading towards a science of qualities which may help in our efforts to reach a more balanced relationship with the other members of our planetary society.[24]

In the last paragraph of the preceding section, I used the phrase 'scientific shift of focus'. This is important because, when I write about a new kind of science, what I am really talking about is a profound shift of focus – just like that described by Polanyi in terms of the shift of focus, or attention, from the internalised, and now unattended-to, particulars to the skilful attention to the whole gestalt and its equally gestaltic contexts. This can be thought of in terms of framing; and, in science, the way the problem is framed seems to be decisive. Framing, of course, is not simply a matter of individual choice, since frames are culturally produced – very often in the form of the culture's myths and guiding metaphors. These, themselves, are strongly derived from our embodied experience, as the work of George Lakoff and Mark Johnson has persuasively argued.[25] As Margaret Boden argues in *The Creative Mind*, and I discuss in Chapter 5, frames (like grammars, for example) constitute the possibility of organised conceptual thinking, but frame-breaking is an important part of human creativity.

The way Darwin framed his theory of evolution, in terms of inheritance, random mutation of inherited characteristics and the usefulness of these in terms of greater adaptation to environment, meant that, once the vehicle of inheritance – the gene – was established, evolutionary biology focused ever more exclusively on genetics. As with the

Copernican heliocentric revolution before it, and the Newtonian development of physics, this developed a scientific view of things which tended to be reductive, atomistic and functionalist. In biology, this meant that it was believed that all science had to understand was how genes worked to replicate themselves. The organism itself was just a vehicle for transporting genes from one generation to the next. As Brian Goodwin has argued, Darwin's frame (as with that of his contemporary scientists, and also present population geneticists such as Richard Dawkins[26]) was strongly, if very quietly and even unconsciously, informed by powerful Christian mythologies of the mortal fallible body and the immortal soul, and by narratives of sin, suffering and redemption.[27] But, as Goodwin insists, *all* science is informed by myths (frames), and this should not be seen as a bad thing. Our myths tell us important things about the nature of our being and our indwelling, and we can certainly learn much from both our own mythologies and from the mythologies of other cultures. I shall return to this in Chapter 3.

The historical interest in genes, however, still does not solve the problem of the evolution of forms (morphogenesis) of organisms. Genes are long strings of molecules which are organised into four bases which appear as sequences along the double helix of DNA. As is well known, DNA can not only replicate itself, but can produce mRNA which is capable of producing proteins (and thus bodies). But DNA itself is inert; in order to effect its quite remarkable work it must exist within a cell in which there is 'an elaborate machinery of other molecules and structures ... that are essential for DNA copying and for the production of mRNA and protein'.[28] In other words, DNA is entirely dependent upon its highly organised context, which is the living cell. This business of *context* (as I argued in Chapter 1) continues to exert effects in the matter of cell development during reproduction. The cell's genetic programme, though, is limited; in Goodwin's words, 'a genetic programme can do no more than specify when and where in the developing embryo particular proteins and other molecules are produced, and we know that molecular composition isn't sufficient to explain physical form'.[29] But neither is the still widely accepted Darwinian theory of random mutation and adaptive fitness – in which there are no potential limits on the forms which might evolve. In practice, and even in very distant ecologies (in Europe, Africa and Australia, for example), millions of years of separate evolution produce some differences (marsupial incubation, for example) but overwhelmingly remarkably similar morphogenetic outcomes. As Jesper Hoffmeyer points out, even the evolution of very different species (dinosaurs and mammals, for example) produces recognisably similar morphological types.[30]

Once a significant shift in the framing of scientific thought has taken place, problems can be addressed in a different way. Obviously, this does not mean an end to laboratory science (as Goodwin's discussions in *How the Leopard Changed its Spots* make abundantly clear); rather, laboratory science takes place within a different framing of the questions, and thus, itself, takes the form of experimental research programmes of a different type, because research questions are framed in a different direction and according to the new perception of a higher level framing. In biology, the problem of morphogenesis (why organisms take the shapes they do) is not answered either by genetics or by natural selection.[31] The way it began to be answered was through the cross-disciplinary observation that, under certain conditions, quite different materials (chemicals in one instance, and algae in another, in the examples cited by Goodwin[32]) produced remarkably similar *patterns* of activity and features of emergence. These developments proved to be the effect of complex systems in which the similar patterns were produced by both positive and negative feedback loops of amplification and dampening. What these patterns showed was that the development of organisms was subject to two important factors: internal formal limitations (the principle of self-similarity in complex systems described in Chapter 1) and external environmental limitations (the principle of systemic *co*-evolution, or 'coupling' in the production of more complex *Umwelts*). In other words, evolution is not truly random, but occurs within a certain kind of complex order: 'Organisms are not random assemblages of working parts, the results of trial and error tinkering by natural selection. They reflect a deep pattern of ordered relationships'.[33] Amazingly, human beings have evolved to be the kinds of creatures who not only can intuit this orderliness in nature, and become skilful in respect to it, but also can, now, begin to describe it. What Goodwin and his biologist colleagues around the world have discovered is, as with Polanyi, that 'there is an inherent rationality to life that makes it intelligible at a much deeper level than functional utility and historical accident'.[34]

In Chapter 1 I described the effects of the discovery of complexity as a biologisation of science in general. I said this in order to emphasise the ways in which, across the sciences, a view is forming of the sort of deep interconnectivity which we, now, easily recognise in ecologies. In fact, the biology of complexity I have sketched out above has been highly dependent upon mathematics and computer modelling (in which it discovers remarkable affinities between biological development and mathematical observations such as the Fibonacci series and the classical Greek idea of the Golden mean), and it would be truer to say that physics, chemistry and biology are all simply moving much closer together in their insights about an active, generative and complex world, As Goodwin remarks, 'physics is becoming more historical and generative while biology becomes more exact and rational'.[35] In general, what one can say is that, across the sciences, the mechanistic and reductive model is being abandoned in favour of the idea of phenomena as *all* deeply interconnected and dynamic, with the result being

> the undeniable consequence ... that the old mechanical view of causation, of external forces acting on inert particles of matter, is dead. Physics now recognises that natural processes cannot be described in these terms and that the phenomena we see in nature are expressions of a deeper reality in which in which apparently separate entities are united in subtle but well-defined ways.[36]

Interestingly, and in the attempt to describe dynamism (and agency) in complex systems, both Goodwin and Bhaskar draw upon philosopher of science Rom Harré's idea of 'causal powers'. This describes the kind of thing which active (rather than passive) determinacy *is*. Causal power is the sort of power anything has in virtue of the *kind* of thing it is. Thus, the causal power of the alga *Acetabularia acetabulum*, when decapitated in a scientific experiment (as described by Goodwin), lies in its morphogenetic capacity to regenerate certain patterns, given an appropriate environment.[37] Similarly, although on an entirely more complex scale, the causal powers of human beings ('doing what comes naturally'[38]) reside in their powers, of imaginative and linguistic representation, for example to inscribe and re-inscribe, and to venture more deeply into (mythologically, artistically, scientifically), the patterns of the world. As with *Acetabularia*, though, environment and causal powers (or what Goodwin calls the 'generative field') matters; human beings have proved highly adaptable, but their flourishing, and their self-organising creativity (autopoiesis), are stunted or deadened by stifling or inhospitable environments:

> The life cycle includes genes, environmental influences, and the generative field in a single process that closes on itself and perpetuates its nature generation after generation. Species of organism are therefore natural kinds, not historical individuals as they are in Darwinism. The members of a species express a particular nature.[39]

These human 'patterns', and the conditions of flourishing indwelling, will be explored in more detail in Chapters 3 and 4.

The point to emphasise here, though, is that, even (or perhaps especially) at the most basic microbial level of life, the living world is a vast interconnected, interdependent web of relations – a complex whole in which emergence produces different strata of beings with generative fields which are complexly interwoven. Just as DNA is only lively in its generative field (the cell), so with all living systems. It makes absolutely no sense at all to privilege the idea of competition (and Marx was right to say that Darwin's theory reproduced the economic ideology of his time); co-operation and mutuality are everywhere in nature – and that includes us. Indeed, there is now a substantial body of biological theory which suggests that, with high connectivity, numerous quite simple relations will produce the emergence of life: 'Life doesn't need DNA to get started; it needs a rich network of facilitating relationships. This is co-operation, mutual support and enrichment'.[40] In Chapter 4, I will describe these 'facilitating relationships' in terms of semiotic relationships.

FROM CELLS TO SOCIETIES

In *How the Leopard Changed its Spots*, Goodwin's detailed discussion of biological science finally moves on to the human stage with his discussion of a science of qualities. Modern science developed as a study of quantities – of the measurable (what Galileo called the 'primary' qualities of objects) features of objects (such as mass, position and velocity) in dynamical relationship with each other. The 'secondary' qualities which were not quantitatively measurable (such as shape, colour and texture) were, as it were, set aside. Biology followed this route with its interest in genes. However, and as the preceding discussion should indicate, while important and revealing information has been obtained in this way, it can tell us nothing about organisms and their generative fields. It cannot tell us much about the conditions of creaturely flourishing. The complexity sciences, however, in understanding the behaviour of complex totalities (from cells to societies), begin to enable us to have a sense of what a science of qualities would be. This involves the frame shift which I have been describing in which what matters is not, say, genes, individuals and cultures seen as more or less 'isolated', and more or less 'static' differences, but a concept of life, and of individuals and societies and cultures, as processual, inter-related and emergent.

This notion of complexity and emergence, and Polanyi's idea of the rationality of nature, is also shared by Jesper Hoffmeyer, whose work I shall discuss in more detail in Chapter 4. For the moment, though, it is worth noting the Cartesian gap which Hoffmeyer also sees as endemic in modern science. Scientists, he reminds us, are not interested in the world as such but in data. Data, however, is not 'in' the world; it is, rather, the source of a collective agreement (amongst scientists) concerning how the world is, but in a limited sense. And it is dependent upon a set of theories and metaphors. It is 'theory loaded', or ideological. Hoffmeyer notes that technological development 'seems to work', but then asks 'How well does it in fact work?', especially when there is much we witness (the placebo effect for a simple example) but still don't understand.[41] The problem is the gap between data and the wider world view, which scientists, too often, are disinclined to think about, considering it 'philosophy' and nothing to do with them since they are merely dispassionately objective collectors of data. But, as Polanyi argues, this is far from the case; and, as Hoffmeyer (very much

2. 'That Eye-on-the-Object Look'   75

not alone) argues, a large part of the problem lies in the deeper Cartesianism underlying modern science. This 'either-or' dualism (rational mind or material body) means that a materialist science cannot allow itself to conceive of rationality in nature:

> Since this rational substance is unacceptable all we are left with is the purely physical. What gets left out of such an analysis is the evolutionary perspective, the idea that a system could be more or less rational; that rationality is something that can occur at levels other than that of the human psyche. If reason does not spring from some special substance but is inherent – rather as a potential within the physical material ... and which a forgetful (mortal) self-referential system might manage to "tell" its descendants about – then this opens up a non-dualistic perspective that has never been tested simply because it conflicts with the dualistic either-or philosophy which not even materialism seems able to shake off.[42]

But a certain rationality does appear in nature with its evident forward directedness: from inert to active, from chemical to biological, from simplicity to complexity, and from what Hoffmeyer describes as semiotic limitation to greater semiotic freedom. This 'from-to' structure is evident at every level of emergence, with each level forming the ontological substrate of the emergent one. One cannot look at evolution with clear eyes and not see this.

Human knowing is participative, and the field of participation is the human generative field of a relational social order. Goodwin quotes the social anthropologist Tim Ingold:

> I have shown how a theory of persons can be encompassed within a more general theory of organisms, without compromising the role of human agency or denying the essential creativity of social life. This creativity, magnified a thousandfold by the work of consciousness, is but a specific aspect of the universal capacity of organisms to act, in a certain sense, as the originators of their own development. It has been said that, in history, 'man makes himself', creating from within the very world in which he is a participant. But man (or woman) is an organism, and organisms generally make themselves, creating as they do a history of life. To arrive at this conception of the organism, however, we need a new biology, or should we say an old one? – for its holistic aspirations are redolent of a pre-Darwinian worldview. It must be a biology that asserts the primacy of processes over events, of relationships over entities, and of development over structure. Organism and person do not then confront one another as specific configurations of matter and mind, 'two sorts of independent substances' as Whitehead put it, 'each qualified by their appropriate passions'. Both are rather embodiments of the total movement of becoming that Whitehead so memorably described as 'a creative advance into novelty'.[43]

The idea that organisms are each of a natural kind, and that each develops according to its own kind (where 'kind' is understood as the organism's generative field), and has causal powers according to its kind, suggests that full flourishing is full activity in all a creature's capacities. This is discussed further, in the context of Michael Marmot's epidemiological research, in Chapter 4. One of the disasters of conceptual thought divorced from experiential thought has lain in the violence it has done to humankind, whether in destroying bonds of trust and mutuality, and ripping up traditional knowledges and ways of life (including, importantly, traditional forms of sociality, as described by both Robert D. Putman and Richard Sennett[44]), or in despoiling generative fields of becoming, and potentialities of various kinds. Human creativity is expressed in play-like ingenuity, but the commodification of labour has robbed people of their own time and, perhaps most importantly, *rhythms* of labour. Of course, modernity provides many compensations in terms of commodities and labour-saving devices; it is also broadly progressive. But its attack on the human spirit, on needs which are *not* met by commodities, has been comprehensive. Modern scepticism has encouraged critical people to think of spiritual needs as inevitably regressive or as a sort of salve which will take the edge off a necessary political fervour. This modern conception of the generative field of humans is much too narrow; the passion felt on a march in support of a political cause is very much the same sort of passion a Christian feels at mass, a follower of Islam praying to Mecca, nearly any human at the sea's edge. The intensity of feelings aroused by political commitments are not different in kind from those aroused by spiritual ones. It is not original to say it, but the fact deserves some proper attention.

Interestingly, Polanyi does not think it was modern scientific scepticism as such which brought about the particular problem which is Enlightenment modernity, but rather this scepticism combined with 'a new passion for moral progress':

> The new social aspirations had their origins in Christianity, but they were evoked by the attacks on Christianity. It was only when the philosophy of Enlightenment had weakened the intellectual authority of the Christian churches that Christian aspirations spilled over into man's secular thoughts, and vastly intensified our moral demands on society.[45]

For him, sceptical lucidity sees that morality in modernity no longer has any anchors in authority or religion, and the moral imperative is turned against itself (in the individual) in the existential scouring out of any remaining 'bad faith', and the declaration of nihilism, and (in politics) in the purism of a marxism in which 'moral perfectionism had no place left for truth':[46]

Moral skepticism and moral perfectionism thus combine to discredit all explicit expressions of morality. We have, then, moral passions filled with contempt for their own ideals. And once they shun their own ideals, moral passions can express themselves only in anti-moralism. Professions of absolute self-assertion, gratuitous crime and perversity, self-hatred and despair, remain as the only defenses against a searing self-suspicion of bad faith. Modern existentialists recognise the Marquis de Sade as the earliest moralist of this kind. Dostoyevsky's Stavrogin in *The Possessed* is its classic representation in fiction. Its theory was perhaps first outlined by Nietzsche in *The Genealogy of Morals*. Rimbaud's *Un Saison en Enfer* was its first major epiphany. Modern literature is replete with its professions.[47]

What Polanyi opposes, of course, with his argument about tacit knowledge is the idea that all knowledge is abstract conceptual knowledge. Combined with moral perfectionism, as I argued in Chapter 1, conceptual knowledge alone, separated from and vastly elevated over experiential knowledge, is capable of producing monstrosities such as Stalinism or the proposition that all human societies should be organised according to market relations to which all social relations should, hence, be subordinate. This latter repellent and inhuman idea is thrust down our throats daily. The commodification of human relations is attempted in laws such as GATS (General Agreement on Tariffs and Services), in which the morality of free access to basic human goods such as healthcare and education becomes transmuted into seeing them solely as commodities in the marketplace. And if nothing as decisive as the 1834 Poor Law is likely to be enacted to commodify relationships based on love and trust, or to turn joy into a commodity (though heaven knows capitalism has tried), then we must ask why this is so.

NOTES

1. Polanyi, *The Tacit Dimension*, op. cit., p3.
2. Polanyi, *The Tacit Dimension*, ibid., p4.
3. I shall return to this. The point to understand is that evolution is not even. As suggested by Stephen Jay Gould and Niles Eldridge's 1972 'punctuated equilibria' thesis, evolution as we understand it thus far consists in long periods of relative stability punctuated by periods of (relatively speaking) very rapid and seemingly chaotic 'revolutionary' episodes in which new species, *of greater biological complexity*, emerge. N. Eldredge & S.J. Gould, 'Punctuated Equilibria: an alternative to phyletic gradualism', in T.J.M. Schopf, ed., *Models in Paleobiology*, San Francisco: Freeman Cooper & Co., 1972.
4. Polanyi, *The Tacit Dimension*, op. cit., p10.
5. Sudnow, *Ways of the Hand: A Rewritten Account*, op. cit.
6. Polanyi, *The Tacit Dimension*, op. cit., p13.
7. Hoffmeyer, *Signs of Meaning*, op. cit.

8. Polanyi., *Tacit Dimension*, op. cit., p15.
9. Think of the example I offered in Chapter 1 of driving skill. Once acquired, the remarkable way in which we internalise the car's size and shape (or extend our mind into it) is capable of adjusting to all sorts of different sizes and shapes of motor vehicles. Do you know how you do this? No, you don't; it's a tacit skill.
10. Polanyi, *The Tacit Dimension*, op. cit., pp16-17.
11. Polanyi, *The Tacit Dimension*, ibid., p18.
12. Hoffmeyer, *Signs of Meaning*, op. cit., p35-6.
13. Polanyi, *The Tacit Dimension*, op. cit., p20.
14. Polanyi, *The Tacit Dimension*, ibid., p22.
15. Polanyi, *The Tacit Dimension*, ibid., p24.
16. Polanyi, *The Tacit Dimension*, ibid., p29.
17. M. Donald, *Origin of the Modern Mind: Three Stages in the Evolution of Culture and Cognition*, Cambridge, MA: Harvard University Press, 1991.
18. See P. Freire, *Pedagogy of the Oppressed*, tr. M. Bergman Ramos, New York: Continuum, 2000.
19. This is clearly demonstrated in the letters between Bryan Magee and Martin Milligan in *Sight Unseen*, op. cit. Milligan was, effectively, blind from birth, and although, deriving his arguments from the use of conceptual knowledge alone, he was able to argue that he could have a good grasp of the world afforded to the sighted, it is quite clear that the absence of vision meant that he could never experience that particular initiation into the visual world which the experience of, and practice in, seeing allows.
20. Keller, *Century of the Gene*, op. cit.; and also J. Hoffmeyer, *Signs of Meaning*, this chapter, note 12 above.
21. D, Bohm, *Wholeness and the Implicate Order*, London: Routledge & Kegan Paul, 1980. The reader will not be surprised to learn that Michael Polanyi and David Bohm were friends in their later years.
22. Polanyi, *The Tacit Dimension*, op. cit., p55.
23. M. Boden, *The Creative Mind*, 2nd edition, London: Routledge, 2004, pp44-6.
24. Goodwin, *How the Leopard Changed its Spots*, op. cit., pxiv.
25. See, for e.g., G. Lakoff & M. Johnson, *Metaphors We Live By*, London: University of Chicago Press, 1980; G. Lakoff & M. Johnson, *Philosophy in the Flesh: The Embodied Mind and Its Challenge to Western Thought*, New York: Basic Books, 1999.
26. R. Dawkins, *The Selfish Gene*, Oxford: OUP, 1976.
27. Goodwin, *How the Leopard Changed its Spots*, op. cit., p28ff.
28. Goodwin, *How the Leopard Changed its Spots*, ibid., p5.
29. Goodwin, *How the Leopard Changed its Spots*, ibid., p40.
30. Hoffmeyer, *Signs of Meaning*, op. cit. p34.
31. Goodwin, *How the Leopard Changed its Spots*, op. cit., p78ff.
32. Goodwin, chapter 3: 'Life, an Excitable Medium', *How the Leopard Changed its Spots*, ibid.
33. Goodwin, *How the Leopard Changed its Spots*, ibid., p98.

34. Goodwin, *How the Leopard Changed its Spots*, ibid., p105.
35. Goodwin, *How the Leopard Changed its Spots*, ibid., p158.
36. Goodwin, *How the Leopard Changed its Spots*, ibid., p160-1.
37. For the alga *Acetabularia*, the crucial environmental factors are calcium levels and temperature in the seas where they live. Outside certain parameters, neither reproduction nor self-repair will occur.
38. Goodwin, *How the Leopard Changed its Spots*, ibid., p159.
39. Goodwin, *How the Leopard Changed its Spots*, ibid., p163.
40. Goodwin, *How the Leopard Changed its Spots*, ibid., p174.
41. Hoffmeyer, *Signs of Meaning*, op. cit., p91.
42. Hoffmeyer, *Signs of Meaning*, ibid., p93.
43. Goodwin, *How the Leopard Changed its Spots*, ibid., p190.
44. R.D. Putnam, *Bowling Alone: The Collapse and Revival of American Community*, London: Simon & Schuster, 2000; R. Sennett, *The Corrosion of Character: The Personal Consequences of Work in the New Capitalism*, London: W.W. Norton & Co., 1998.
45. Polanyi, *The Tacit Dimension*, op. cit., p57.
46. Polanyi, *The Tacit Dimension*, ibid., p60.
47. Polanyi, *The Tacit Dimension*, ibid., pp58-9.

CHAPTER 3

Intimations: nested enfoldings – stages in knowing

Before moving on to a discussion, in Chapter 4, of the 'writing' on our bodies in terms of health and mortality that can be discovered in contemporary research in epidemiology, and also to an attempt to theorise this within a fuller biological account via the new field of biosemiotics as found in the work of Jesper Hoffmeyer, I want to take what, to some, may seem to be a detour through a discussion of some aspects of religion and the arts. This 'detour', though, seems to me to be an important one. When I asked, at the end of the last chapter, why is it that love and joy seem to be resistant to commodification, I meant to raise the question of human experiences of investment and pleasure that are not, in any easy sense, reducible to rational utilitarian calculation by numbers, but might be understood in a more integrated, and even scientific, way. In the last chapter, I looked at some accounts of science from within a systems perspective, that refuse a reductively objectivist version of scientific knowing. In what follows in this chapter, I look at some further accounts of affective life that allow us to think about what they have in common with a scientific way of thinking.

My reason for writing this chapter is not to slip into, or naïvely to avow, an unpolitical mysticism as an escape from the difficult realities of science and politics, but, rather, to explore the relationships between these different ways of knowing in order, hopefully, better to understand them. The view that reason and emotion, the mind and the body, are best kept apart is a result of a long history of philosophical thinking in which rationality – reasonable thinking and behaviour – was seen as a way of escaping from brutal conflict. We must, surely, still hold to and affirm this view. But while this idea eventually led to a rejection of primitive superstitions, beliefs, and the unassailable right of powerful groups or individuals to impose their will on others by force, our contemporary understandings of the intimate relationship between mind and body, reason and passion, has been considerably extended in the past ten to fifteen years.[1] We can no longer say, if we are to remain properly scientific, that affective life is something simply to be disregarded, but, instead, must come to understand it better. Politics, itself, is not to be progressed by ignoring the affective intimations of the millions of our fellow beings who find both solace and joy in the appearance of order in the universe, who express this in forms of sacred belief or the wish that natural environments be preserved for solace as well as sense, or who insist that life, in all its forms, is valuable in and of itself. The purpose of this chapter, then, is to try for a more integrated way of thinking about those three great spheres of human activity: science, morality and art. In this, I will hold off from theorising in the main – although I will look at the ways in which General Systems Theory's idea of subjectivity as located and processual is reflected in Buddhist philosophy. What I will be suggesting, though (quietly following Sebeok), is that all forms of human knowing (pre-articulate, mythic, modern and postmodern in John Deely's sense) are forms of 'world modelling', which imply, also, conceptual 'framing'. In Chapter 5, I will take this discussion further in the direction of theory by looking at the ways in which human creativity can be understood as the manipulation, and sometimes breaking or transforming, of models and frames, and the narratives (what Margaret Boden calls 'scripts') which they generate.

The kinds of pleasures experienced in art and religion are much older than supposedly 'cool', supposedly 'rational', ways of knowing things, and religion and art form a substrate to science in the modern sense. But, equally, I have just argued in the preceding chapter, largely by way of Michael Polanyi, that modern science is in some way continuous with older ways of knowing, and just as passionate. Indeed, if we revert to the older sense of the word 'science' as simply knowledge, we can see that modern science, and then complexity science, simply build upon their parent knowledges. As Robert Laughlin argues, 'nature is now revealed to be an enormous tower of truths, each descending from its parent, and then transcending that parent, as the scale of measurement increases'.[2] But what this means is that we do not (as has sometimes been thought) dispense with older ways of knowing (although modern science has certainly involved wholesale disavowals of them – in the Freudian sense of unhappy and incomplete repression). These 'older' ways – narratives and passions, let us say – remain within, or as the supporting storeys, of the towers whose further levels, while seeming quite different in design, are nonetheless architecturally dependent upon them.

The late twentieth and early twenty-first century version of the Speenhamland Laws (i.e. the interim arrangements which effectively held up the commodification of labour until the 1834 Poor Law), standing between us and the supposedly hyper-rationalist logic of capitalist commodification of affective life is, surely, the vast amount of scientific work concerned to show that human feelings are not irrational things to be eradicated from mature and progressive thought, but, on the contrary, are absolutely essential to human reasoning.[3] Equally, our increasing idea of the importance of emotional intelligence and emotional literacy, and our growing emphasis on the quality of our affective relations and of relationship time, let alone our knowledge of the affective effects of our behaviour upon children, all suggest important advances, both in human self-understanding and in resistance to the kind of utilitarianism which nearly drove poor John Stuart Mill mad, until he read Wordsworth and discovered that he wasn't simply a Benthamite calculating machine after all.

With the development of complexity science, we understand better our connectedness and co-arising being, and also the widely felt sense that many things which do not seem immediately causal (in 'old' science terms) actually are. There is now perfectly substantial research which backs up our strongly felt hunches and intimations about the general harmfulness of social fragmentation and rationalisation; of, for example, great extremes of wealth and poverty within any one society. Whether you take as your source the neurobiology of Antonio Damasio, the philosophy of Michael Polanyi, or the realism of the hunches, instincts, intimations and so on that are given weight in Michael Marmot's epidemiological work, the idea that feelings are important sources of real knowledge is gaining ground. Damasio's research in *Descartes' Error*, for example, demonstrates that feelings (which are the form in which intimations arise in us) are an essential part of rational behaviour. When the neural pathways between the (evolutionarily ancient) amygdala (the site of our affective responses) and the language processing neo-cortex are damaged, but the latter isn't, people can calculate and calculate, but they can't make choices, and their sociality is blighted. That we *do*, indeed, 'know more than we can tell', and that this knowing very often proceeds by ways of hunches, instincts and so on, seems to be indubitable. That this knowing, on the negative side, can also result in apparently inexplicable neurotic or otherwise pathological behaviour is also very obviously the case. I will discuss this in Chapter 4. A more general discussion of hunches and clues in the context of creativity will follow in Chapter 5.

Once you take a complex systems view of things, the move from effect to cause is clearly anything but simple; and shared *intimations* (what Brian Goodwin calls 'a science of qualities') must be treated as potentially *real* sources of knowledge in complex systems. At least one implication of this, furthered by Candace Pert and team's work on the activity and distribution of neuropeptides – rich, of course, in the brain, but also, and interestingly, in the gut – is that we should give more attention than we habitually do to hunches, gut feelings, and so on.[4] A short story recounted by Robert Winston in his 2003 BBC1 series *The Human Mind* will serve as an example here. Winston interviewed a fire-fighter, the leader of his team, on an amazing escape from a blaze phenomenon called blow-back. The team were fighting a bad blaze in a warehouse. They had begun to bring the blaze under control and had entered the building. Things were looking quite good when this team leader suddenly had a strong premonition of danger which caused him to order his men out of the building. There was no apparent reason for this, and some of the other fire-fighters began to demur. But, no, said the team leader, do as I say and get out of the building now. Sensible of the proper chain of command, the men withdrew. Almost immediately, there was a huge explosion and fireball caused by the violence of the fire which, having consumed all the air within the space, had now sucked in, with great force, the fire from adjoining parts of the building. The team were saved because their team-leader had listened to what we might call his gut instincts. In fact, Winston conjectured, the man's experience had filed all his memories of other fires he had been in, one of which included precisely this rather rare form of fire behaviour known as blow-back. He had not consciously retrieved these memories; consciously, he was busy fighting just that particular fire. But what his unconscious mind gave him was a very urgent affective response tied to this one unconscious memory of the features of fires under certain conditions. Paying attention to a feeling had saved his own and his men's lives.

Intimations are, of course, the stuff of art, literature and religion. They furnish us with the interior life of other minds (and times and cultures), and with the phenomenologically textured experience, in the case of art and religion, of ritual, invocation, and the represented invocation of sensuous experience in material forms. In what follows, I want to explore what seem to me to be a related set of intimations: human distress at the loss of (the immediacy of) the natural world; a sense of the sacred – especially as indwelling in nature; and a way of thinking about the relationship between the experience of the sacred, nature and art, and science.

NATURE AND LOSS
Clearly, where our modern experience of nature is concerned, the unprecedented thing – in terms of a decisive break with all earlier and other forms of human lived experience – is the eighteen-century Industrial Revolution which grew up in Britain on the back of the scientific revolution begun in the previous two centuries. Nature's presence in the context of some form of troubling technological innovation can be seen as early as 1740 in James Thompson's *The Seasons*. Awareness of the Industrial Revolution's shadow, and fear of its effects, is written all over the culture of late eighteenth- and early nineteenth-century England – and, of course, continues to be the central problem around which all other questions revolve throughout and beyond the nineteenth century. What was being lost was what Ford Madox Ford

called 'the quiet thing' associated with rural England and the value of locatedness and stillness. In Jane Austen's *Mansfield Park* (written between 1811 and 1813), the impending threat is repeated again and again in the contrast between Fanny Price's stillness – her utterly insistent passivity – and the dangerous restlessness (and recklessness) of Henry and Mary Crawford. The latter represent even a new kind of self: charming, exciting, but superficial, 'acted' and driven by money and material goods, without any sense either of religious values or of the older rural culture's feudalistic sense of the ways in which material things are tied up in reciprocal obligations.

Even as relatively 'late' as 1854-5, in Elizabeth Gaskell's *North and South*, the Milton-Northern (Manchester) mill owner John Thornton is presented as a tempestuous Byron of industry, and the striking mill workers as sublimely inchoate. There is clearly something inhuman about technological industry even at this time. In *North and South*, Southern England represents a peaceful hierarchical order which is already untenably slow and intellectually backward (one thinks of the 'midnight darkness' of the countryman Dagley's mind in Eliot's *Middlemarch*), whereas Milton Northern represents energy and a more emancipated future (Margaret Hale is struck by the jaunty independence of the young female mill-workers). Even so, and whilst Margaret's pastoral associations are altered by her move to Milton, it is Thornton who must become encultured by an education in the classics, and it is Margaret's new fortune – the swelling of her power – which, in marriage, 'saves' Thornton's half-industrialised soul. Gaskell's arrangement between the forces of city and countryside is typical of the 1850s and 1860s hope (Matthew Arnold is the representative figure here) that education – in poetry, no longer the Classics for the swelling middle classes – and the gentler values of pastoral would tame both industry and political economy alike. In fact, the traffic was rather in the other direction.

This loosening of social bonds and speeding up of an increasingly mechanised life is noted as deracinating and potentially deforming by most contemporary observers – with more than one commenting on the assault on the senses produced by the sheer complexity and productive activity of the new cities (especially London and Manchester), and (by the 1830s) the speed of the new steam trains. Some note the ways in which the poor city dwellers themselves begin to reproduce this restlessness as if by choice and, even when offered the possibility of security, seem more inclined to a life of uncertainty. Philip Davis notes the case of a fish-hawker woman as reported by a stipendiary magistrate of Thames Police Office to the *First Report of the Royal Commission on the Poor Laws* (1834). This woman almost certainly earned enough to provide a stable condition of life, but 'preferred an alteration of great privation and profligate enjoyment'. Davis goes on to say that although, for the magistrate, this was an example of 'immoral improvidence', to Engels, who had noted the case in the *Report*, 'the apparently perverse behaviour was an externalised form of thinking, symptomatically miming at its own level the speculative risk and gamble of the wider economic world itself'.[5] Certainly, one is reminded of the forced ups and downs of Moll Flanders' life in Defoe's novel over one hundred years earlier in the newly expanding commercial world of early Georgian England. Wordsworth, similarly, in the 1802 Preface to *Lyrical Ballads*, notices the disturbed condition of London's citizens, 'where the uniformity of their occupations produces a craving for extraordinary incident'.[6]

But amongst Davis's wide range of examples in his wonderfully expansive and informative literary study *The Victorians*, perhaps the most telling example of the sometimes intolerable new experience of up-rooting movement is his description of the peasant poet John Clare's move from the tenement he shared in Helpston, Northborough (now Northamptonshire) with his parents, his wife and their six children, to the larger cottage just three miles away provided in 1832 by his patron the Earl of Fitzwilliam. Confronted with this advantageous move, Clare was barely able to tolerate it. Davis quotes 'The Flitting' where Clare writes:

> I've left my own old home of homes
> Green fields and every pleasant place
> The summer like a stranger comes
> I pause and hardly know her face.[7]

The trauma was such, Davis notes, that Clare's own sense of self was broken with it; driven insane from the unrootedness of a mere three miles, he was soon to be removed to an asylum.

But the fear of mechanisation and the consequent deformation of humanity reported by so many nineteenth-century contemporaries (Wordsworth, Carlyle, Dickens, Arnold, Ruskin, Morris, for example), and noted by virtually all subsequent commentators (G.M. Young, Walter Houghton, Philip Davis, for example) in which the very structures of thought and feeling were being changed, did not recede in the twentieth century. Weber, Durkheim, the Frankfurt School and, more recently, psychoanalytically informed writers such as Christopher Lasch, have all been troubled by the denuding of experience, and the incapacity to express such alienation – to find words adequate to modern experience – as identified by Walter Benjamin in 'The Storyteller' after the First World War.[8] In his 1983 novel *Waterland*, Graham Swift describes his narrator's father, Henry Crick, once famous for his Fenland tales, returning from the mechanised hell of the Great War silent, with no stories left to tell. In his 1996 *The Spell of the Sensuous*, which details the author's experience of refinding his proper sensate life far from so-called civilisation in the jungles of the Far East, philosopher David Abram tells again the story of this cumulative human loss.[9]

What we must surely say about this great weight of historical and on-going contemporary evidence from so many disparate mouths is that it is a wholly insufficient response to mark it off as simply romantic and nostalgic. Romanticism, as I argued in Chapter 1, is, anyway, itself a complex response to technologising modernity, and far too important to turn into a dismissive cliché. As I have argued elsewhere,[10] any human response (or figuration) that is (and has been) so widespread and so pervasive over time demands a serious analysis. To refuse such, and to write it off as 'conservative' or 'reactionary' as some have wanted to do, is to refuse dignity to the reporting of human experience – fallible as that can sometimes be. The time for entertaining these time-lagged philosophies bounded by nearly three hundred years of Western history dogged by idealist mistakes is surely past. As George Lakoff says in *Women, Fire, and Dangerous Things*, the findings of contemporary phenomenological cognitive science indicate that our understandings of human being and reasoning 'are now in transition'.[11] It is in all these ways that we have re-encountered the question of nature in our own time and in the growth of green (and red-green) politics.

Nature, as Raymond Williams pointed out, is a word of great complexity – one capable of the most diverse definitions.[12] And, as Kate Soper notes in *What is Nature?*,[13] it has been the subject of much dispute between postmodernists and realists, i.e. between those who claim that nature is a discursive construction, too often used as a way of normatively defining ideologically what is 'un-natural', and those of a materialist, phenomenological and ecological persuasion who insist that nature is real (and certainly as real as culture), and a source of intrinsic value both in itself and also for humans.

In what follows here, I am mainly interested in looking at the ways in which nature is (in a Polanyian sense) both the stuff which science attempts to apprehend, and, at the same time, that which appears ordinarily to humans as beautiful, strange, patterned, powerful and something about which *more can be known*. This latter sense approaches the world with a personal feeling of what we often describe as the sacred; though Polanyi's argument, as I hope I have shown in Chapter 2, is that this passion is, secretly, what animates scientific research also. It is for this reason that, as John Pickering amongst others has pointed out, science in modernity seems so often (and as much as scientists themselves may dislike the fact) to have taken the place of religion as a source of authoritative statements about the world.[14] But it seems to me that, beneath the ideology of stern objectivity, the actual *doing* of science (which I discuss in the section entitled 'Doing Science, Joyful Indwelling' below) shares a quite close kinship with both a 'religious' experience of surrender to the world and with the ordinary joy of being in the natural world which is such a common experience. The experience of surrender as immersion in, as a form of giving one's self over to experience – whether a scientific problem, an environment, a book, a painting, a meditation, a prayer, or a piece of music – is, of course, a process of active semiosis in which I do not simply receive signs, but actively embrace them. A learning or skilful mind is akin to a pair of hands which grasps, holds, and turns about. When the object is a made thing, this active surrender is also intersubjective. As Merleau-Ponty says:

> Language continuously reminds me that the 'incomparable monster' which I am when silent can, through speech, be brought into the presence of *another myself*, who re-creates every word I say and sustains me in reality as well ... Speaking and listening not only presuppose thought but – even more essential, for it is practically the foundation of thought – the capacity to allow oneself to be pulled down and rebuilt again by the other person before one, by others who may come along, and in principle by anyone.[15]

And what is true of speech is true of all intersubjective semiosis.

It is, though, in its elevation of disembodied, abstract conceptual thought, and in its occlusion of personal, embodied and passionate knowing, and embodied tacit skills in knowing and indwelling, that modern science still misleads both us and itself. It is also this false idea of objectivity which has led increasing numbers of scientists (with perhaps psychologists predominant) to turn to non-western ways of conceiving the world and living skilfully in it.[16] As Pickering notes, psychology, above all human sciences, is about subjective experience, and yet contemporary mainstream cognitive science has found no way of approaching the subjective being in which it is so interested:

> But while cognitive neuroscience may yield finer and finer descriptions of the spatio-temporal structure of brain activity, what is this going to tell us? This is a description of the vehicle for consciousness, not of the dynamic flow of consciousness itself. It is to miss the emergent causal powers of the larger system in which brain activity is participating.[17]

Here, again, what psychologists such as Pickering, Varela, Thompson and Rosch have found in Buddhist meditative techniques is a holistic attitude to knowing, and understanding of phenomenological embodied knowledge, which is participative:

> What is required ... is an appreciation of how mental life is embodied and enacted, that is, how it participates in generative patterns of biological and cultural action. This participatory view points directly to the troublesome issue of experience. Troublesome, that is, for psychology

in an era when science has lapsed into scientism. The treatment of experience, or rather, the exclusion of experience, was what Husserl depicted in the *Crisis* [*of European Sciences and Transcendental Phenomenology*] as a misdirection of Western psychology. Although Husserl's critique was directed at the behaviourist psychology of his day, it applies *mutatis* to cognitivism. In the years since the *Crisis*, the murmur that Husserl detected at the heart of psychology has become clearer. Furthermore, during this period the devaluation of the human condition and massive environmental damage both stand revealed as major consequences of scientific technocracy. The disquiet over the exclusion of experience is a part of the growing dissatisfaction with an unrestrained science that seems to contribute to human alienation rather than to ease it. Heidegger presents this as a consequence of decoupling science from human values.[18]

We can see from all this why the question of nature, and the ways that we know ourselves in the world, is tied both to an idea of sacred knowledge as participative and to a feeling for nature, and our skilful being in the world, itself.

DOING SCIENCE, JOYFUL INDWELLING

> The religion of the future will be a cosmic religion. It should transcend a personal God and avoid dogmas and theology. Covering both the natural and the spiritual, it should be based on a religious sense arising from the experience of all things, natural and spiritual as a meaningful unity. If there is any religion that would cope with modern scientific needs, it would be Buddhism.
>
> Albert Einstein

In fact, and before going on to a discussion of the unacknowledged roots of skilfulness in the world as an aspect of the sort of knowing which humans experience as something akin to *sacred* experience, I want suggest that what we experience as sacred is, precisely, derived from the (to modern science) mysterious nature of our skilfulness in the rationality of nature. To be sure, this rationality is not that abstracted conceptual thing which modern philosophy has called rational, but to say this is merely to notice the slippage of a word which is, appropriately, as complex as the word nature.

In furthering my argument here, and for the purpose of making things difficult by an appeal to the experience which we have discovered is so 'troublesome' (as Pickering puts it), I want to talk about an experience which I think will be familiar to many of my readers. I do so as a way of advancing the suggestion I made above when I said that doing science (i.e. research), as with non-scientific research, shares a close kinship with the kind of surrender to the mystery and rationality of the world that William James, for example, describes as religious experience; and both these share a kinship with that pleasurable and skilful indwelling that we experience in the natural world. This argument is more or less explicit in Polanyi's *Personal Knowledge*, the final chapter of which makes the same argument concerning evolution, natural selection and morphology as is found in Goodwin.[19] At the chapter's close, Polanyi describes both evolutionary progress, and the scientist's passionate drive towards more knowledge of an ordered universe, in terms which identify these two kinds of activities as an unfolding 'towards an unthinkable consummation. And that is also, I believe, how a Christian is placed when worshipping God'.[20] Tellingly, the experience I am about to describe is also a good description of human creativity, and its themes, of hunches, hints and clues, will be picked up again in Chapter 5.

When, as academics, we do research, we really *don't* know (as Polanyi avers) what will arise. We have a sense of a real problem, but we really don't know how it will be solved. This is akin to what happens in speaking and writing generally in which, most usually, we don't know precisely what we are going to say until we say it. It is also akin to what Husserl says in the *Crisis*. When we are doctoral students, this is an enormous problem, because we are not yet usually sufficiently skilled in our discipline to break free from learned rules in order to 'go for the jazz' as David Sudnow puts it.[21] We cast round for other skilful expositions which will allows us to say at least a little of what we want to say. When we are more skilful (have internalised a wider body of experience and become more experienced ourselves), we have intuitions about fields which are more likely to be useful (or 'fruitful', as Polanyi would say). But we still don't know how knowledge, and some possible answer to our problem, will proceed.

According to the procedures of our particular discipline, we write proposals; but we know these won't really describe the *actual* processes of our encounter with rationality – because this will derive from tacit skills that we can't 'propose' in conventional linguistic form. Then we read books, or while away our time with other relevant pursuits such as programmes of laboratory or field research, but what we are *actually* doing, no matter how well we try to disguise this fact from ourselves, is *having faith* in the rationality of the problem we have discerned, and *waiting* for a way forward to present itself to us. *Of course*, the books, or the scientific literature, or the fieldwork is very important, because it furnishes us with the particulars with which we add (if we are already skilful) to our stock of conceptual and tacit knowledges. In this way we make our environment rich and fertile. But the move to the next stratum, if it comes, arrives in a relatively mysterious way, over time, necessarily (and hence, presumably, the gain to be had from sleeping on a problem) as we are able, unconsciously, to move to dis-attending from the particulars to a focus on the gestalt whole.

The essential creativity of a research programme is found *not* in all the books we read (though these will certainly help us, by prompting us to different thoughts and authoritative insights), and *not* in the experiments we devise (though these experiences – the semantic root of experiment and experience is the same – will certainly help us in proximal ways), but is to be found in our faith that something real will come through. Although it is, officially, anathema to modern scholarly research (because 'alarmingly' subjective), the truth is that the closer we can bring our research questions to our own lived and skilful experience, the more likely these are to yield good results. Candace Pert, for example, describes her scientific research in precisely these terms.[22] But this lived and skilful experience is tacit; we know more than we can tell. Thus, when we are really skilful, we follow leads which are essentially hunch-like. The more we give ourselves over to an apparent serendipity, the closer (paradoxically according to modern theories of knowledge) we come to real discovery. There is, thus, an apparent (and real) randomness to research and creativity, but, as Polanyi argues, randomness is a feature of emergence.[23] We simply *don't know* all the complex particulars via which new strata emerge. If we did, then randomness would not be present and no new strata of knowledge would or could emerge. It is only when we shift our attention from the internalised particulars (of many experiments conducted; of many books read) and allow that a certain kind of randomness must be present, that our impassioned driving on into things can produce something new. This may be easier to see, and to acknowledge, in artistic creativity and in research in the humanities, but it is, in fact, a characteristic of all investigation. It is, in fact, characteristic of *all* human creativity and culture. Very skilful researchers in any discipline have an intimation of this and are willing to have the kind of faith which allows them to surrender to this ordered and rational – but not orderly or explicitly rationalisable – process. With this, we can say that science is an art which, at its most skilful, is also characterised by faith in the nature of the world and its rational unfolding. I shall have something more to say about this is my concluding Chapter 5 where I return briefly to the question of creativity and biosemiotics.

In his *Varieties of Religious Experience*, William James writes that, while the stoical consciousness accepts reality, and very often determines to act morally within that personal acceptance, the tone of this stoical acceptance is noticeably 'cold'. Stoicism is a philosophy of self-mastery born in slavery. In contrast, he adds, the tone which distinguishes a religious acceptance of the world is personal, passionate and joyful. In other words, both submit, or surrender, as it were, but the tone of personal religious (as opposed to stoical) submission (which is not at all the same as rule-following submission to a doctrine) is a passionate and pleasurable, 'warm', surrender. This, it strikes me, is very like the intensely pleasurable surrender to the mysterious actions of what I must describe as the 'becoming of tacit knowing' which the skilful researcher (whether scientist or humanist) experiences.

In the 'official' modern world of modern science and the modern metaphysics which subtends it, this is heresy. But as countless reports attest, this is, in fact, how *all* creativity (including scientific creativity) proceeds.[24] We might also note that, again like language, what is going on is actually a collective affair – even though books, and now the world wide web, very often have to stand in for the conferences, colloquia, symposia which we recognise as such an important part of a scholarly community. And perhaps we might see the racking-up and intensification of demands for precision about what we do as researchers, and the auditing explanations of our every human move, as late modernity's increasingly desperate search (or demand) for the principles of conceptual order upon which that modernity has believed itself to depend. In the now manic intensity of the disciplining demand (very often framed in terms of the need to respond to rational consumer choice – another metaphysical fantasy – in the marketplace), we can perhaps see the dawning recognition of the horrible fact that human knowledge and behaviour is not calculable in this way, and that the way that humans do produce knowledge (and manifestly they do) is, in the end, utterly elusive and opaque to that particular fantasy of reason. The increase in contemporary demands for orderliness and calculability can, thus, be understood as one of the 'morbid symptoms' of the (post-modern) phase-shift constituting the interregnum between one understanding of what constitutes knowledge and a new one.

SCIENCE AND THE BODY

Of course the idea of nature can be thought of as a cultural construct inasmuch as different cultures, and all cultures at any point in history, have various ideas about what nature is. But it remains the case that the natural world (however, culturally or historically, we define it) is the ground-state of our being. It is so because we are embodied and enworlded creatures. Our minds would not be the common, shared and shareable, things they are were this not this the case – although the metaphysics of individualism generally wishes to deny the obvious case that subjectivity is intersubjectivity, and that human being indwells. Indeed, phenomenological objections to the expulsion of the body and experiential knowledge in Western metaphysics can also be seen as an objection to the ways in which such a metaphysics attempts to preserve the human-nature, subject-object, distinction itself. Similarly, various religious expressions of the distinction between the pure and the impure, in which, first expelled as impure, the latter then returns harnessed as 'an apt symbol' of great generative power,[25] can be seen to be concerned with the same problems of boundaries and distinctions.

As we shall see, Buddhism deals with this problem in quite a different way. But in many religious traditions, including Christianity, the expulsion of the impure body and its return as sacred object (the abjected body of Christ on the cross, for example) is testimony to a deep anxiety. This essentially religious anxiety is repeated in modern science. In *What is Nature?*, where she considers the ways in which nature has been used in philosophy and religion, and attempts to establish a realist ontological basis for the valuation of nature, Kate Soper points to phenomenological challenges to such Western metaphysics:

> In a broad sense, we are here talking of a materialist challenge to the puritanism and elitism of the idealist refusal of the body. Though this has been registered in materialist approaches in philosophy from very early on, it is not until the nineteenth century that it is given powerful and sustained expression – notably in the argument of Nietzsche, Marx and Freud. Thus Nietzsche exposes the repudiation of the body as a morbid moralisation through which humanity learns to be ashamed of its own instincts, and invites us to read the history of philosophy as 'an interpretation of the body, and a misunderstanding of the body'. Philosophy, he suggests, can be read as narrative about the body that takes the form of a denial of its philosophical validity, since it says 'away with the body, this wretched *idée fixe* of the senses, infected with all the faults of logic that exist, refuted, even impossible, although it be impudent enough to pose as if it were real'.[26]

Thus we can see that western metaphysics, including its expression in modern science, performs the same gesture in relation to purity and the body that religious ritual performs. It expels it, and then what is expelled returns – significantly at the end of the nineteenth century, when the modern idea of science as a qualitatively different sort of knowledge has gained firm ground – as a source of tremendous generative power in the philosophy of phenomenology. I also noted in Chapter 1 that this gesture marks the beginning of the modern period, in which the repression of the medieval carnivalesque body is begun in the development of courtly manners and protocols. The complexity and inter-relatedness of the medieval world picture returns, albeit at a different stratum, upon modern science (indeed is hatched within it, according to the same logic of the carnivalesque ritual inversion and celebration of the 'earthy' body) in the form of complexity science, and possibly as its apotheosis.

It should be clear from the above that while modern science, by the end of the nineteenth century, has marked its endeavour off as something qualitatively different from art and religion, in fact it is not. This is not to denigrate science – it is a tremendously productive, powerful, and sometimes glorious, human activity – but is to indicate that, just as with the evolutionary nature of the relationship between nature and culture, science is evolutionary, cultural, and marked by just the same human gestures and concerns as are art and religion. It does not stand above its human origins as an oracular God to whom all must be permitted; it is recognisably human through and through.

Given this confraternity, two (at least) things follow. One is that the flat claim that something can't be true or valuable because it 'isn't scientific' can no longer be made in the absolutist way it has often been done in modern societies. As Polanyi, discussing emergence and knowledge, puts it:

> This kind of emergence is known to us from inside. We experience intellectual growth in the process of education and, in more dramatic forms, in the creative acts of the mind. I may recall in particular the process of scientific discovery. This process is not specifiable in terms of strict rules, for it involves a modification of the existing interpretative framework. It crosses a heuristic gap and causes thereby a self-modification of the intelligence achieving discovery. In the absence of any formal procedures on which the discoverer could rely, he is guided by his intimations of a hidden knowledge. He senses the proximity of something unknown and strives passionately towards it. Where great originality is at work in science or, even more clearly, in artistic creation, the innovating mind sets itself new standards more satisfying to itself, and modifies itself by the process of innovation so as to become more satisfying to itself in the light of these self-set standards. Yet all the time the creative mind is searching for something believed to be real; which, being real, will – when discovered – be entitled to claim universal validity. Such are the acts by which man improves his own mind; such the steps by which our noosphere was brought into existence. For in the ontogenesis of the innovator we meet a step in the phylogenesis of the human mind.[27]

From this (and indeed from everything else I have argued here thus far) it must follow that claims to scientific objectivity as legitimation of any activity, especially perhaps political activity, should be seen for the ideological sham that they are. Science is a wonderful thing. Indeed, it is so wonderful that, in its developments in complexity science over the past fifty years, it has shown itself to be fully and wholly a human thing, full of humanity's own causal powers and marked by human ways of going about things.

Second, embedded in its relational human being, science, nonetheless, and as powerful a tool as it is, cannot claim that it is of an entirely different order of knowing to knowledge in art, the humanities and religion. It should follow from this that technocratic and instrumentalist claims based upon the supposed superiority of scientific and technological knowledge should be seen as exercising no especial prerogative over claims from humanities or arts disciplines. As Ilya Prigogine hopes, the 'two cultures' divide should gradually erode, and with it all claims to the *intrinsic* superiority of scientific ways of knowing.

With this, we should not find it surprising that frankly religious forms of knowledge begin to return within science – especially where that has been informed by the development of complexity science. The rediscovery of rationality, or order, in nature, and especially the discovery of chaos, bifurcation and spontaneous self organisation (autopoiesis) in all open systems, returns us to a powerful sense of mystery and pattern, and also to the powerful wisdom expressed in sacred understandings, and in their more secular versions in ecological sensibilities where nature is experienced as expressive of a particular order and source of wonder for humans (and possibly for other animals also). Indeed, and especially in play, *all* animals can be seen to express a joy in their lively explorations of skilful being in their own generative field.

Once again, it's probably worth remembering that, until the nineteenth century, the Italian word *scienza* (from which the English 'science' is derived) simply means knowledge. Prior to the development of modern science, religion and philosophy were the expression of human attempts to understand the world, and to live meaningfully and well in it. They *are* the science of the pre-modern world. Similarly, and as John Pickering points out, Buddhism 'arose in a cultural context where the opposition of science and religion did not exist. The Abrahamic religions of the West are based on belief and revelation while Buddhism is not. It is not a faith but a system of mental culture based upon direct investigation of the lived world'.[28]

Buddhism is an atheistic philosophy of lived experience, and, we might say, a much older form of highly developed phenomenology. Indeed, Husserl was familiar with Buddhist philosophy, and, as Pickering notes, 'From Schopenhauer onwards, explicit use has been made of Buddhism by Western philosophers, most notably Heidegger'.[29] With the development of complexity science, the philosophy of which is phenomenological, that use seems likely to continue. In *Mutual Causality in Buddhism and General Systems Theory*, Joanna Macy describes the compatibility of Buddhist insights with open systems theory,[30] and several psychologists, as mentioned in Chapter 2, have also found in Buddhism useful sources for thinking about human subjectivity.[31] Roy Bhaskar's recent work, alarmingly to the many marxist-orientated followers of Critical Realism, has also turned to Eastern philosophical religious traditions as a source of further understanding. The point is, of course, that, once the confraternity of science with other forms of knowledge has been recognised, there is no point in continuing to think that religious, or artistic, or humanities thought has nothing interesting and real to say about reality. Indeed, marxism itself, which has long inspired the kinds of passionate commitments found in religious belief, might well cease to think of itself as differently defined as a science, since science, itself, can no longer be considered to be anything other than a particularly potent form of *belief* and *praxis* in the rationality of nature.

NATURE AND RELIGION

> 'Wonderful, lord, marvellous, lord, is the depth of this causal law and how deep it appears. And yet I reckon it as ever so plain.' 'Say not so, Ananda, say not so! Deep indeed is this causal law, and deep indeed it appears. It is through not knowing, not understanding, not penetrating, that doctrine, that this generation has become entangled like a ball of string, and covered with blight, like unto munja grass and rushes, unable to overpass the doom of the waste, the woeful way, the downfall, the constant faring on.'[32]

Of course, religions are human ways of trying to understand the world and to live well (according to their generative field, we might say, or in semiotic freedom) in it. As Mircea Eliade argues, sacred intimations are rooted in human experience of the natural world. Here, again, the emphasis falls on human experience of the sacred as the breaking up of homogenous space, which is without focus and thus fraught with relativity, by the introduction of sacred space in which the sacred space (as with Mecca for Muslims, or Jerusalem on the Mappa Mundi of the medieval world) creates an orientation.[33] This orientation is also an opening, or threshold, between the world of man and the world of the gods. It is thus, Eliade argues, more *real* than the profane space which surrounds it, and full of real power: 'Thus it is easy to understand that religious man deeply desires *to be*, to participate in *reality*, to be saturated with power':[34]

> The sacred is pre-eminently the *real*, at once power, efficacy, the source of life and fecundity. Religious man's desire to live *in the sacred* is in fact equivalent to his desire to take up his abode in objective reality, not to let himself be paralysed by the never-ceasing relativity of purely subjective experiences, to live in a real and effective world, and not in an illusion.[35]

Understood in this way, science is the continuation, and evolution of, an ancient impulse to articulate reality in the forms of communally shared understanding. The sacred, it follows, comes into being with language, and the development of metaphoric abstraction, as a way of anchoring and confirming experience *as shared* – as 'objectively' real. It is the enactment, in culture, which marks what, in the next chapter, I will discuss in terms of a conceptual membrane. It is, thus, unsurprising that we are willing to accord culture, and thus some form of language, to those vestiges of hominid life we find marked by the use of burial and grave goods. Science, we might say, is the most powerful way of doing magic (which thereby loses its purely pejorative connotations) that we have thus far discovered. For Eliade, modern humans never throw off entirely their ancient sacred myths and understandings. Thus, to cite just two examples amongst many possible ones, the sacred experience of a new home as the 'founding of a new world', in imitation of the original world-founding of the gods in the struggle with chaos, remains still in the celebration event of the house-warming; similarly, the importance of homologous world-foundings and losings such as births and deaths are also celebrated or otherwise marked – even if only in the token form of flowers and cards. Our passions, as we now know, arise from an evolutionarily ancient part of the brain (the limbic system), and we have not lost them – nor could we ever.

Inasmuch as the sacred impulse arises from the passionate desire to live in, to participate in, reality, it is the same desire which animates scientific knowledge. And if the civilisation of manners is, as Elias argues, the precondition of modern science, because of the idea of objectivity and 'objective reality' this allows, we may also say that modern science construes itself, similarly to sacred impulsions, on the basis of the urge not to be 'paralysed by the never-ceasing relativity of purely subjective experiences' and 'to live in a real and effective world, and not in an illusion'. In other words, scientific knowledge is contiguous with sacred knowledge and with the desire ever to reach further towards the 'more to be known'. The problem with this incredibly efficacious new religion lies in its belief that it must (or can) denude itself of the world of experiential knowledge and the body.

A word of caution is probably necessary at this point. In arguing that the impulse to the sacred lies very deep within us, and in saying that science partakes of this as much as any other human activity does, and that both religion and science are about collective affirmations of shared 'objective' reality, I am, of course, conscious of the ways in which, when formulated into organised religions, this impulse can be regressive, defensive and destructive. As a source and guarantor of community – especially when the latter is perceived to be under economic threat – the passions of religion (and this may to some extent be true of scientific communities also[36]) often take socially unacceptable forms. In asserting the depth of the lineage of sacred intimations and their relationship to the affirmation of community in its experiential bonds with nature and place (especially sacred places), I merely wish to point out that, far from being an escape, as is sometimes thought, from rational political action and modernity, a community organised around its ties to such deep intimations is really 'doing politics' by other means. All of which is simply to say that the organisation of religion – as an organisation of ancient and powerful intimations – is always political, and that, in modernity, religion irrupts as 'politics' when modern politics according to liberal ideas of justice, have failed.

Certain forms of religious fundamentalism may be overtly opposed to modernity, but they are also a product of it.

For Eliade, psychoanalysis is an example of mythic thinking and initiation rites resembling the descent into the underworld (here, the unconscious) rendered in secular form.[37] But, if psychoanalysis can be said to contain elements of the sacred (which I think it can), this is not at all to say that this observation means that psychoanalysis is thereby fraudulent or inefficacious, any more than the observation that modern science is contiguous with religion should be seen as saying that science is fraudulent or inefficacious. The whole point of what I am arguing here is that modern, scientific, ways of knowing more about nature (including, of course, about human beings) are not *essentially* different from pre-modern ways of knowing; but they are problematic to the extent that, in trying to live in an 'objective reality', they try to forget what is *not* forgotten, because it is actually *lived*, in sacred apprehensions of the world, i.e. the experiential, skilful and tacit, basis of all human knowing.

In accepting that psychoanalysis retains elements of a sacralised world-view, but is nonetheless (and at its best) both efficacious *and* a general advance in human self-understanding – especially, perhaps, in its recognition of the importance of 'signs written on the body' – we can, perhaps, usefully integrate its 'secular' insights with those 'religious' ones offered by Buddhism – and then also compare each of these with the definite increase in human self- and world-understanding offered by the development of General Systems (or complexity) Theory. Does the sophistication of Buddhist philosophy, in understanding the inter-relatedness and co-dependent co-arising of all life (*paticca samuppada*) which complexity science now also apprehends, offer us any further clues about human life within such a context? I think it does, especially in the ideas of mindfulness practice, right action (or right livelihood), desire- or craving- induced suffering, no self, and impermanence.

When Buddhism says that everything is impermanence, and that there is no such thing as a self, this does not mean that things are not real. Rather it means that, to use a more Western term, everything is real but is *process*. We know, for example, that no human being who had never been in the company of another human being would be able to develop (for itself) the humanity which we recognise in each other. Human animals do sometimes, very rarely, survive in the absence of human community, but when they do, their subjectivity is not much more than that of the animals in whose company they have survived. Human subjectivity, on the other hand, is intersubjective; it is the result of our social nature and our social being. It is, in other words, the result of communicative processes and other minds and bodies. A human mind not only results from, but continues to be, the process of being passed through other minds (and arms). As Joanna Macy writes:

The systematic principles von Bertalanffy discovered in biological process extended, he maintained, to the person *in toto* as a psycho-physical entity. As such the person is an irreducible and dynamic whole, in open interaction with her world, sustaining and organizing herself through appropriation, transformation and differentiation of meanings and symbols.[38]

Identity is emergent, processual and always reaching forward towards the 'more to understand' which is discovered, as knowledge, in *scienza*. As I have argued above, in the Western metaphysics which, in Western culture, divides science from other ways of knowing such as art or religion, this reaching forward has been reified in conceptual knowledge divorced from experiential knowledge, and objectified in science; but, clearly, people do still also do this reaching forward in religious or quasi-religious terms – and, indeed, as Polanyi argues, science also reaches forward in impassioned ways. Nonetheless, in the *misconceived* materialism of our epoch, this reaching out very often takes crudely materialistic and acquisitive forms – whether this is the acquisition of property, or people as property, and so on. It is this latter form of reaching forward into life which produces suffering.

There is, though, another form of impassioned reaching forward towards the 'more to understand' which is not harmful. This form of reaching forward is one that understands that everything is real, but also understands that what is real is processual, and can only be fruitful when what it has in mind is *the process* itself. This is mindful practice, and it means that, *because everything is real* (including motivations, intentions, effects on the well-being of others, deeds, ideas, means – the *whole* of our existence and the existence of all living and inanimate things), fruitful living can only be achieved through right action (or right livelihood). In other words, intentions, motivations, ideas and so on, can only be really fruitful (for us and others) when they are in the service of a genuine attempt to further 'the path', which simply means the *process* of self-understanding of the world. If I am mindful of the processual self that I am, and mindful of the subtlety of my co-arising, co-dependent processual existence among other creatures, and mindful that this means that *all* my doings reach out into all life, my reaching out will be spontaneously directed in right ways which are not primarily motivated by narrow self-interest. This is not a matter of some mystical fantasy of being 'good', or 'spiritually pure', by renouncing self-interest and conscientiously going about 'caring for others, or the Other'; it is about recognising the processual nature of self (as constituted) in community, and pursuing the furtherance of, and real experience of joy in, that. It is a full expression of what Amartya Sen calls human 'capabilities', and Jesper Hoffmeyer calls semiotic freedom. I discuss this further in the chapter which follows, where I look in more detail at the epidemiological evidence concerning what happens to people when this processual nature of self in community is thwarted.

I would say that the truth of this deep causality, and the fruitfulness which comes from trusting in it (or having faith in it), is, retrospectively, observable. The closer an individual comes to the mindful living of it in the present, the greater the experience of joy in its experiential happening or unfolding. But, really, this is nothing different from the skilfulness in being (the kind of knack for surrendering to the rationality of the world which is found in the structure of tacit knowing) that Polanyi describes as the actual process of doing science, and which I have described above in terms of the ways in which research actually happens. One could say that, at its best, when conceptual knowledge is not divorced from experiential knowledge, and is, thus, able to give itself over to the tacit, experiential, structure, *this* is what science *is*. It is a personal, experiential, informed commitment to something universal, impersonal and beyond mere self, and a faith in the reality of the 'more to be known' in nature. Paradoxically, when science understands what it is *really* doing when it is at its most successful, it exemplifies Buddhist practice. In other words, it is processual; intuits that there is a problem and, thus, that there is 'more to be known'; doesn't know what this is (i.e. is not instrumental); but proceeds with the passionate conviction that more truth will, indeed, eventually unfold itself.

MUTUAL CAUSALITY

Complex systems involve multiple feedback loops (positive and amplifying, and negative and tending to stabilise) and emergent behaviour. Although we live in an epoch in which the apparent discreteness of our bodies (and their hidden away bodily functions) encourages the fiction of monadic individualism, in fact our sociality is systemic and relational. It should be clear why this sense of gestalts is of interest to psychologists, because it is the whole relational system which needs to be understood (whether family, organisation, crowd, city), not simply individual parts – which are close to meaningless when considered in such an isolated way. Similarly meaningless (to the point of unintelligibility) is the view that consciousness is a 'spandrel' (a functionless epiphenomenal development), since it is human consciousness which allows the possibility of communication (information flows) and feedback via which the gestalt is maintained and can evolve.

Clearly, the understanding that subjectivity is relational was one of Freud's major insights (the psycho-semiotic nature of the biological symptom was another); in the psychoanalytic relationship the analyst hopes to be loved (emotionally invested in, which is the meaning of transference love) and also to become the enigmatic screen upon which will be played out all the relationships which are there, constituting the analysand, but not physically present in the consulting room. Essentially correct also, but better thought about in terms of bodily or gestural 'information', is the Freudian idea of the symptom as something 'to be read'. Here, pathology, whether directed against the self in the form of illness, or against others in the form of covert aggression, arises as meaningful precisely because of our generalised skills in human being. This is tacit knowledge of the deepest kind in which we know something (that certain kinds of symptoms or symptom-acts are 'information'), but don't *know* that we know it.

Merlin Donald's argument (mentioned in Chapter1 and discussed further in Chapter 5) about the gestural prehistory of spoken language suggests that, with the achievement of spoken language and conceptual thought, this kind of awareness of embodied and gestural information remained tacit and unconscious.[39] The apparently high level of resistance that humans have to acknowledging the kind of embodied information manifest in psychopathology (although all humans 'send' and 'receive' this kind of embodied signalling all the time) may well stem from its belonging to an earlier, closer to the animal, stage in our evolutionary history.

As both Buddhism and a systems informed psychology understand, self is not something fixed, but is process. The self is like a journey (or a path), and what an individual *is* at any point on that path *is* simply the path, or, in other words, the network or web of relationships – the environment in all senses of the word – in which they are enmeshed. And, of course, that path has a specific history, but it is not predetermined; it is always in the process of becoming: 'as Buckminster Fuller says, "I seem to be a verb"';[40] and, as Norbert Wiener writes:

> It is the pattern maintained by this homeostasis which is the touchstone of our personal identity ... We are but whirlpools in a river of ever-flowing water. We are not stuff that abides, but patterns that perpetuate themselves.[41]

We are, as Brian Goodwin puts it, and like every other living thing, creatures with a generative field, and our processual life cannot be thought about coherently without understanding this. Joanne Macy quotes Ervin Laszlo's 1969 *System, Structure and Experience* to make the same point:

> We must do away with the subject-object distinction in analyzing experience. This does not mean that we reject the concepts of organism and environment, as handed down to us by natural science. It only means that we conceive of experience as linking organism and environment in a continuous chain of events, from which we cannot, without arbitrariness, abstract an entity called 'organism' and another called 'environment'. The organism is continuous with its environment, and its

experience refers to a series of transactions constituting the organism-environment continuum.[42]

Notice the semiotic element captured in the notion of 'transactions'.

ECOCRITICISM AND ECOSEMIOTICS
At the end of his 2003 book *After Theory*, Terry Eagleton writes:

> We can never be 'after theory', in the sense that there can be no reflective human life without it. We can simply run out of particular styles of thinking, as our situation changes. With the launch of a new global narrative of capitalism, along with the so-called war on terror, it may well be that the style of thinking known as postmodernism is now approaching an end. It was, after all, the theory which assured us that grand narratives were a thing of the past. Perhaps we will be able to see it, in retrospect, as one of the little narratives of which it was so fond. This, however, presents cultural theory with a fresh challenge. If it is to engage with an ambitious global history, it must have answerable resources of its own, equal in depth and scope to the situation it confronts. It cannot afford simply to keep recounting the same narratives of class, race and gender, indispensable as these topics are. It needs to chance its arm, break out of a rather stifling orthodoxy and explore new topics, not least those of which it has so far been unreasonably shy. This book has been an opening move in that inquiry.[43]

In the last decade of the twentieth century, the only obviously new critical formations in cultural studies to 'break out of a rather stifling orthodoxy and explore new topics, not least those of which it has so far been unreasonably shy', and to bridge the 'two cultures' gap between the arts, forms of 'sacred' experience (especially, of course, in relation to nature), and the sciences, have been biosemiotics and ecocriticism. Still new,[44] and drawing on a family of traditions (nature writing and Transcendentalist philosophy in the USA; Romanticism, ecology, cultural materialism, and complexity theory in the UK) rather than a developed single methodology and body of thought, ecocriticism, encompassing, as it does, for example, ecological science, ecojustice, literature and phenomenological philosophy, may well offer a way forward from the relative stagnation which Eagleton identifies. Certainly, ecocriticism's interest in the human preoccupation with the natural world which arises in philosophical and literary Romanticism as a reaction to the experienced threat of loss of this world, and its ecological interest in 'the organism-environment continuum', draws it into the need to bridge the divide between both literary concerns and contemporary scientific ideas – especially, and obviously, in biology. It is too soon to tell whether ecocriticism will manage to extend itself into a comprehensive theory. Its main body is the Association for the Study of Literature and the Environment (ASLE) which was set up in the USA, and affiliated to the Modern Languages Association (MLA), in 1992, and in the UK (ASLE-UK) in 1996. Subsequent ASLEs have formed in Europe, Japan, India, and Australia-New Zealand. Terry Gifford, of ASLE-UK writes:

> As a relatively new movement in cultural studies, ecocriticism has been remarkably free of theoretical infighting. There have been debates about emphasis and omission, but these have not directly challenged the positions of the originators of the movement. They have rather pointed to new directions for research into ecofeminism, toxic texts, urban nature, Darwinism, ethnic literatures, environmental justice and virtual environments, for example. Ecocriticism has not developed a methodology, although its emphasis on interdisciplinarity assumes that the humanities and science should be in dialogue and that its debates should be informed equally by critical and creative activity. These are radical enough practices for those located within the disciplinary and career demarcations of the academy.[45]

Its common ground is certainly a concern for the natural world and ecological conservation, but, to date, it seems very open to a wide interpretation of the idea of what constitutes an environment, with the latter, for all the interest in natural environments, including both the quality of the environment wherever that is found, and also concern for human built environments as well as plant and animal ones.

A related development is the growth, mainly a Northern European phenomenon among biologists, of ecosemiotics, which, again, and like its relative biosemiotics, is concerned with producing an integrated account of the biosphere (or semiosphere) in which humans, and human language and culture, are not seen as fundamentally different from the rest of the world, but where the idea of semiosis is found to be common to all life. Indeed, as the semiotician Thomas Sebeok has written, the definition of life may coincide with the definition of semiosis. The hypotheses of biosemiotics as described by biologist Kalevi Kull are: 'semiosis arose together with life; semiosis, symbiosis, and life processes are almost identical (or isomorphic); life is mainly a semiotic phenomenon, the real elements of life are signs'.[46] Similarly, in *Signs of Meaning in the Universe*, Jesper Hoffmeyer hopes that some of the problems which biosemiotics can solve are: 'to reformulate the concept of information; to transcend (overcome) the dualism of mind and matter, i.e. the mind-body problem; to solve the incompatibility of humanities and natural sciences; to unite cultural history to natural history; to give humanity its place in nature'.[47]

Biosemiotics and ecosemiotics both draw on the same sources: the biology of Jakob von Uexküll (1864-1944), who first used the term *Umwelt* to describe any organism's 'experience' of its environment as

3. Intimations

an active space of semiosis; the semiotics of C.S. Peirce, who considered the entire universe to be 'perfused by signs'; and the synthesis of these by the semiotician Thomas Sebeok. For von Uexküll, the environment (*Umwelt*) is not the 'outer world' described by Ernst Haeckel (who, on the basis of his understanding of Darwin, coined the term 'ecology'), but is, rather, the organism's subjective world. As Winfried Nöth puts it:

> *Umwelt*, in this sense, is the way in which the environment is represented to the organism's mind, and it comprises the scope of the organism's operational interaction with its environment. Because of the species-specific differences between organisms, their different needs, capacities, and perspectives of their environment, there are as many kinds of *Umwelt* as there are species (or even organisms). Every species and every organism can only perceive whatever the biological structure of its receptors, its brain, and its specific perspective of its environment allows it to perceive.[48]

An organism's *Umwelt* is, therefore, what *signifies* for it; and through it it perceives stimuli and responds to them. An *Umwelt*, in other words, is a space of semiosis.

I will give a more detailed account of biosemiotics, and the developed synthesis of biosemiotics and complexity theory offered by Jesper Hoffmeyer, in the chapter which follows, in which I turn to a discussion, or the beginning of a discussion, about the ways in which complex systems and emergence

NOTES

1. A. Damasio, *Descartes Error: Emotion, Reason and the Human Brain*, London: Picador, 1995.
2. Laughlin, *A Different Universe*, op.cit. p208.
3. Damasio, *Descartes' Error*, op. cit.
4. See W. Wheeler, chapter 5, *A New Modernity? Change in Science, Literature and Politics*, Lawrence & Wishart, 1999; Hoffmeyer, chapter 6, *Signs of Meaning*, op. cit.
5. P. Davis, The Oxford English Literary History, Vol. 8. 1830-1880, *The Victorians*, Oxford: OUP, 2002, p36.
6. W. Wordsworth, 1802 Preface to *Lyrical Ballads* in Wordsworth and Coleridge, *Lyrical Ballads*, eds. R.L. Brett and A.R. Jones, London: Methuen, 1968, p249.
7. Davis, *The Victorians*, op. cit., pp33-4.
8. W. Benjamin, 'The Storyteller', *Illuminations*, tr. H. Zohn, New York: Schocken Books, 1968.
9. D. Abram, *The Spell of the Sensuous*, New York: Vintage, 1997.
10. W. Wheeler, 'Nostalgia Isn't Nasty', in M. Perryman, ed., *Altered States: Postmodernism, Politics, Culture*, London: Lawrence & Wishart, 1994.

11. G. Lakoff, *Women, Fire, and Dangerous Things: What categories reveal about the Mind*, London: University of Chicago Press, 1987, pxii.
12. R. Williams, *Keywords*, London: Fontana Press, 1988.
13. K. Soper, *What is Nature?*, Oxford: Blackwell, 1995.
14. J. Pickering, 'Selfhood as Process', in J. Pickering, (ed.), *The Authority of Experience: Essays on Buddhism and Psychology*, ed., J. Pickering, Richmond: Curzon, 1997, p153.
15. M. Merleau-Ponty, 'Science and the Experience of Expression', *The Prose of the World*, Evanston: Northwestern University Press, 1973, pp19-20.
16. F.J. Varela, E. Thompson, E. Rosch, *The Embodied Mind: Cognitive Science and Human Experience*, London: MIT Press, 1993.
17. Pickering, (ed.), Editor's Foreword, *The Authority of Experience: Essays on Buddhism and Psychology*, op. cit., pxiii.
18. Pickering, 'Selfhood is a Process', in J. Pickering, (ed.), *The Authority of Experience*, ibid., p156.
19. Goodwin, *How the Leopard Changed its Spots*, op. cit.
20. Polanyi, *Personal Knowledge*, op. cit., p405.
21. Sudnow, *Ways of the Hand: A Rewritten Account*, op. cit.
22. Pert, *Molecules of Emotion*, op. cit.
23. Polanyi, *Personal Knowledge* op. cit., p390ff.
24. A. Koestler, *The Act of Creation*, London: Picador, 1975 [first published 1964].
25. Soper, *What is Nature?*, op. cit., p90, quoting M. Douglas, *Purity and Danger*, Routledge, Kegan Paul, London, 1978, p161.
26. Soper, *What is Nature?*, ibid., pp91-2.
27. Polanyi, *Personal Knowledge*, op. cit., pp395-6.
28. Pickering, 'Selfhood is a Process', *The Authority of Experience*, op. cit., p158.
29. Pickering, 'Selfhood is a Process', *The Authority of Experience*, op. cit., p158.
30. Macy, *Mutual Causality*, op. cit.
31. Macy, *Mutual Causality*, op. cit.; Pickering (ed.), *Authority*, op. cit.,; F.J. Varela, E. Thompson, E. Rosch, *The Embodied Mind: Cognitive Science and Human Experience*, London: MIT Press, 1993; see also my 'The Theo-Ontological Expansion of Science', *New Formations*, 50, Summer 2003.
32. Cited in Macy, *Mutual Causality*, op. cit., p25.
33. M. Éliade, *The Sacred and the Profane: The Nature of Religion*, tr. W.R. Trask, London: Harcourt Inc., 1959.
34. Eliade, *Sacred and Profane*, ibid., p13.
35. Eliade, *Sacred and Profane*, ibid., p28.
36. Robert Laughlin has some trenchant things to say about the ways in which scientific communities organize themselves around economic imperatives and threats. See Laughlin, Chapter 13: 'Principles of Life', *A Different Universe*, op. cit.
37. Eliade, *Sacred and Profane*, op. cit., p208.

38. Macy, *Mutual Causality*, op. cit., p80.
39. Donald, *Origins of the Modern Mind*, op. cit.
40. Quoted in Macy, *Mutual Causality*, op. cit., p111.
41. Macy, *Mutual Causality*, ibid.
42. Cited in Macy, *Mutual Causality*, ibid.
43. T. Eagleton, *After Theory*, London: Allen Lane, 2003, pp221-2.
44. The Association for the Study of Literature and the Environment (ASLE) was first accredited in the USA, as affiliated to the Modern Language Association of America, in 1992, and held its first conference at Fort Collins in 1995; ASLE-UK was set up in 1996, and subsequent ASLEs – in Europe, Australia, India and Japan – have been organised in the first decade of the twenty-first century.
45. T. Gifford, 'Recent Critiques of Ecocriticism', *New Formations*, (forthcoming)
46. K. Kull, 'On semiosis, Umwelt, and semiosphere', *Semiotica*, vol. 120 (3/4), 1998, pp299-310. This article can be read on-line at www.zbi.ee/-kalevi/jesphohp.htmz. The quotation to which this note refers, and the following quotation referred to at note 43, are from p6 of the on-line version of the article.
47. Kull, 'On Semiosis', ibid.
48. W. Nöth, 'Ecosemiotics', *Sign Systems Studies*, 26, 1998, pp332-343. This article is available on-line at http://www.ut.ee/SOSE/noeth.htm. The quotation to which this note refers is on p4 of the on-line version of the article.

CHAPTER 4

Perfused with signs: biosemiotics and human sociality

> To modern science, dualism still holds good as a way of dividing the world into two kingdoms, those of the mind and of matter, the cultural and the natural spheres. Non-intervention is still the easiest compromise and one which ensures that both the humanities and the natural sciences can get on with their work undisturbed. And it is this boundary that biosemiotics seeks to cross in hopes of establishing a link between the two alienated sides of our existence – to give humanity its place in nature.[1]

In the preceding chapter, I argued that religion, art and science are not fundamentally different sorts of human endeavour, but must be understood as related activities and emergent features in the complex evolution of 'towers of truths'.[2] In Chapter 3, I described these as 'world-modelling' systems. I also tried to show some of the ways in which these apparently different concerns can be understood in relation to each other. I ended the chapter with the observation that literary and cultural studies come together with the biological sciences in the new cultural studies development of ecocriticism. I also implied that the development of biosemiotics might be understood, itself, as an 'ecocritical' development, and as a way of providing a unified, or integrated, field in which something previously understood as the preserve of humans alone (semiosis) was actually a feature of all living things. In this chapter, I will move from the observations of biological complexity in general – which are found in the work of Brian Goodwin on 'generative fields', and of Humberto Maturana and Francisco Varela on 'coupling' – to a specifically human account of stratification, emergence, and tacit knowledge as that can be explored and perhaps a little better understood in the theory of biosemiotics further developed by Jesper Hoffmeyer – via the epidemiological research of Michael Marmot, the psychoneuroimmunology described by Paul Martin, and (very briefly) the sociological work of Robert D. Putnam.

In *The Tree of Knowledge: The Biological Roots of Human Understanding*, Humberto Maturana and Francisco Varela make a detailed biological argument (similar to Brian Goodwin's) from what *must* logically be the case in the emergence and evolution of life – from the first single-celled organisms right through to human beings – that a sensorium is never, and cannot ever be, properly understood as separable from its environment.[3] An organism, a living thing, is structurally determined as an autopoietic, i.e. self-organising, thing subjected to 'drift', i.e. evolution (although not in the passive Darwinian sense), over time. But its ontology, and its drift, Maturana and Varela argue, is the result of structural coupling of a first order: with its environment; and then, of a second order, with the development of multi-cellular life, as cells combine and evolve new morphologies. The first order structural coupling (organism and environment) still remains or persists (and we see this as every new generation phylogenetically repeats this history, beginning, as always, as a single cell); the second order describes emergent combinations, which Jesper Hoffmeyer describes, when evolved into insect and animal life, in terms of 'swarms' and 'swarm intelligence'. We see this, literally, in the lives of complex systems such as termite nests and bee colonies, but the idea of swarm intelligence can also be used to describe the combined working of new morphologies into very complex creatures.

This works as a good description of complex biological life; but it also works at levels of emergence ('towers') beyond that, in the emergence of societies. We know that biologically simple entities (ants, for example) which perform according to relatively simple biosemiotic 'rules' (here pheromone messages where simple iteration produces complex behaviour[4]), if deprived of contact with the nest, wander aimlessly and quickly die. More complex entities (humans, for example), which are themselves 'swarms of swarms',[5] and operate according to complex semiosis and in much greater 'semiotic freedom' (which sustains them, through language, memory and writing, for much longer), can, accordingly, tolerate isolation much better and for longer. Nonetheless, it is no coincidence that solitary confinement is considered a serious form of punishment amongst humans, and that severely isolated humans eventually succumb to symptoms of psychosis and sometimes suicidal despair: 'No man is an island'.

As with Brian Goodwin's arguments discussed in Chapter 2, Mircea Eliade's and Joanna Macy's arguments discussed in Chapter 3, and Maturana and Varela's arguments alluded to above, it simply makes no experiential or biological sense to talk simply about 'individuals', but only about the organism-environment continuum. Of course, individuals are real, and have 'minds', but these minds (from the simplest to the most complex) are more usefully understood in terms of semiotic processes which necessarily and logically include the environment in which an organism swims – its 'world' or *Umwelt*. When we consider the development, or evolution, of more complex forms of life, we are also looking at the development of both greater biosemiotic complexity and greater biosemiotic freedom. With the development of language, culture and society, human beings are the most biosemiotically free creatures to have evolved so far. However, and as I have already argued and shall come on to discuss further, the evolutionary structure of complexity, stratification and emergence – which Polanyi has described in the formulation of tacit knowledge as 'disattending from in order to attend to' – can also mean that 'disattending from' earlier levels involves a 'necessary forgetting' in the attainment of skilfulness. Specifically, human linguistic skilfulness (the over-emphasis on abstract conceptual thinking in our understanding of the world) can lead to a forgetting, or at least a serious underestimation, of non-linguistic (conscious and unconscious) semiosis. This, I suggest, provides us with the beginning, at least, of a way of using Hoffmeyer's biosemiotic theory as a way of interpreting evidence from both epidemiology and psychoneuroimmunology about the relationship between status – the ability to freely exercise capabilities – and individual occurrences of illness.

For the moment, what we can say is that the world is not simply out there as 'information' to be processed by our senses (the representational view of mind); our human nervous/endocrine/immune system (sensorium) actively contributes to the world it thus calls forth. The human inner world (*Innenwelt*) is in constant dialogue (negative and positive feedback in the complex systems sense) with the human natural and social outer world (*Umwelt*). Thus, we can never talk about absolute objectivity (or absolute subjectivity), but always only about more or less accurate accounts (this is what science tries to achieve) of the triggers and perturbations between complex coupled systems. While these triggers and perturbations are congruent with the maintenance of the systems involved, we can talk about the ongoing 'throbbing of all life'.[6] Where the perturbations are excessive (radical changes in the environment for example – and, where human beings are concerned, we have excellent semiotic reasons for including the cultural, linguistic and affective environment in this account), life may become impossible, and illness or death result. People generally have no problem with the idea that an overly perturbed, i.e. hostile, natural environment can damage or kill; but the idea that a social environment can really make you sick, or even prematurely kill you, is something that, in the residues of a culture of mechanism and individualism, we still find it much harder to believe. In fact, the evidence that stress, including social impoverishment, affects health and mortality is overwhelming – as is shown both in Marmot's epidemiology and in Martin's psychoneuroimmunological explanations of this.

So although unified systems are structurally determined, the relations between systems are not determined; they operate within the limits that describe perturbations and triggers as tolerable or intolerable, life-continuing and life-enhancing or life-denying and life-impoverishing. The tasks of sociology, anthropology and politics consist, at base, in describing (and in the case of politics responding to) those limits for human beings. But they fail to do this adequately when they focus on individuals rather than, properly, on third order (i.e. social) structural coupling. For human beings are comprised of all three orders.

These insights remind us, of course, of the quote from Raymond Williams with which I began this book: 'the consciousness is part of the reality, and the reality is part of the consciousness, in the whole process of our living organisation'. As Williams goes on to note, this is also the insight of Coleridge in the latter's distinction between 'substantial knowledge' and 'abstract knowledge':

> Coleridge spoke of 'substantial knowledge' as 'the intuition of things which arises when we possess ourselves as one with the whole'. This realization, the capacity for 'substantial knowledge', is the highest form of human organization, though the process it succeeds in grasping is the common form of our ordinary living. At a less organized level, we fall back on what Coleridge called 'abstract knowledge, when we 'think of ourselves as separated beings, and place nature in antithesis to the mind, as object to subject, thing to thought, death to life'. The antithesis of nature to the mind, 'as object to subject', we now know to be false, yet so much of our thinking is based on it that to grasp the substantial unity, the sense of a whole process, is to begin a long and difficult revolution in the mind.[7]

We can see from this that when we talk about the environment called forth by the human sensorium, we are always talking not only about the important world of nature, but also about the world of human culture and society. Our natural and cultural worlds *are* us. And, as social creatures, so are other people. Our senses of ourselves, and the extent to which we flourish, are not located primarily in our intrinsic individual selves and capacities, as the philosophy of individualism supposes; on the contrary, they derive from the richness of our social connectedness, from our sense of our importance to other human beings, and from the amount of control we can exercise in our lives – especially over stressful situations and events. The more useful way of thinking about all these things – so as not to confuse control with rigidity or being overbearing, for instance – is in terms of what Hoffmeyer calls 'semiotic freedom'. Since both he and others consider that semiosis is the most fundamental indicator of life (biosemiotics), we can think of a flourishing human life in the same way (although not the same degree) that we think about any kind of animal or vegetative flourishing. For humans, who live in a biosphere which is also a highly articulated semiosphere, maximal flourishing consists in the richness of our semiosis: our contacts, our ability to be heard and responded to, our sense of being supported and effective in a rich number of ways.

Semiotics is a wide field. It encompasses much more than spoken or written signs, or other organised gestural languages such as American Deaf Language. As Thomas A. Sebeok says: 'All known living organisms communicate exclusively by nonverbal messages with the sole exception of some members of the species *Homo sapiens*, who are capable of communicating, simultaneously or in turn, by both nonverbal and verbal means'.[8] A definition of life is very probably simply that which is semiotic. Human communication is very rich in both verbal and nonverbal (conscious and unconscious) semiosis. The tendency, in some quarters, to focus only on abstract articulate language (and to assume that it is essentially rational and unaffected by the emotions) when considering the ways that humans know what they know has led to some serious failures in a full consideration of conscious and unconscious human meaning-making. We can begin to correct this by a better understanding of the field of semiotics as that has developed in the past thirty or so years.[9] A particularly rewarding field in regard to the study of human flourishing and human sociality is that of biosemiotics, in which the developing understanding of evolutionary open adaptive complex systems of stratification and emergence, combined with both evolutionary biology and with semiotics, is producing – particularly in the work of Jesper Hoffmeyer (which I will come on to discuss a little later) – a very rich account of the human life-world, or *Umwelt*.

Before considering this in more detail, I want to draw on a related sphere of biological study, that of epidemiology, to begin to understand the ways in which our very being as individuals, measured in the conditions of our flourishing or failing to, is, in highly significant ways, an effect of our *social* being. Once we have viewed the evidence, though, we will need a way of explaining it. Psychoneuroimmunology can help us with the details; but biosemiotics provides a way of thinking for a fuller account and a new, different and wider way of thinking about what we have discovered; in biosemiotics both biology and language are redescribed so as to bring together mind, body and environment within a new conceptual framework, which moves beyond what Kalevi Kull has described as 'the metaphysical idea that cells and organisms are simply organised organic molecules'.[10] Not only will this help us to think about the coupled relationship between individuals and their environments in ways which emphasise human sociality, it may also help us to begin to think about human natural and cultural environments more widely.

EPIDEMIOLOGY: SIGNS ON THE BODY

In Chapter 1 I touched upon the work of the epidemiologist Richard Wilkinson. Now I want to turn to his colleague, Michael Marmot, and to the latter's thirty years of study on the relationship between illness, mortality and status. It is here that we see most starkly the 'language beneath language', the language of the group, and the ways in which it is written in the fortunes of bodies. Of course, it should come as no surprise that it is in medicine – a science most definitely *not* confined to the laboratory, but very much out there in the real complex world of human experience – that we should encounter the need for explicit or implicit ideas of semiosis. Hippocrates, whom, Sebeok tells us, 'historians have sometimes reverentially regard as "the father and master of all semiotics"', found it necessary to distinguish between 'natural' and 'conventional signs'.[11] My aim is to offer a biosemiotic interpretation of Marmot's evidence of the ways in which both cultural (i.e. overwhelmingly conventional) and natural signs (e.g. the array of signs of human sociality itself) can be written on bodies in the form of natural signs (disease in this case); in doing so it will be necessary to offer a careful account of both these kinds of signs, and to find a way of suggesting how they can, indeed, become apparently continuous.

Philosophical objections to liberal individualism are well known; as, for example, in the work of Charles Taylor and of Alisdair MacIntyre.[12] For MacIntyre, the entire modern language of morality and obligation has become meaningless because we no longer have the pre-modern contexts which once provided the meaning and the force of such arguments. In the absence of these contexts, moral argument is reduced either to the interminable arguments of rationalism, in which perfectly good arguments can all too often be put for quite opposing cases, or to philosophical emotivism in which an affirmation of the goodness of something is simply a linguistic statement incapable of any extra-linguistic grounding.[13] MacIntyre finds no escape from this problem in phenomenology because the latter (he believes) can, similarly, discover no grounds for 'ought' in 'is'. This is because, while phenomenological intentionality certainly grounds consciousness in the body-in-the-world, it cannot easily move from embodied being and consciousness to moral evaluation. The meaningfulness of the idea of a morality, or an ethos, in pre-modern societies (and their concomitant meaningless in ours) can be seen, MacIntyre argues, in the original meaning of (Greek) ethos (Latin translation: *moralis*) – as living and behaving appropriately according to your place in a hierarchical social order, and one, moreover, decreed by God or the gods.

In fact, and as I shall argue below, the significance of 'place', social hierarchy and social connectedness remains very important in human life. Our modern problem lies in recognising this, and its effects, in a way that (in the absence of divine decree) allows us to understand the relationship between social hierarchy and social connectedness on the one hand, and its effects in human biology and physical and emotional flourishing on the other. That there is such a relationship is very evidently manifested in the research evidence which shows that physical and mental flourishing (health or wholeness) is very closely tied to place in social hierarchies; and that this is intimately connected with the level of an individual's social connectedness, and the extent of their opportunities for exercising a wide range of capabilities (relationships, influences, relative autonomy and so on) – or, in other words, their semiotic freedom. In this way it becomes possible to see that socialist and liberal moral arguments concerning the desirability of equality (i.e. the recognition of hierarchy and the need to 'flatten' it) can be satisfactorily grounded in a science of embodied being and consciousness. Retrieving God (or the gods) and a vocabulary of the sacred might be one way of authorising contemporary claims for the equality of our creatureliness; thinking in terms of biology and semiotics (biosemiotics) might be another, perhaps better, one. Equally, though, we might want to note that being in a supportive community is very good for your health, and sociological evidence such as that detailed in Robert D. Putnam's *Bowling Alone: The Collapse and Revival of American Community*, suggests that belonging to a church (as with other kinds of informal communion) is one way of doing this.[14]

As far as complexity in human societies is concerned, some of the most compelling social totality research is to be found in epidemiological studies such as those reported in Michael Marmot's *Status Syndrome*.[15] Marmot's work is not explicitly biosemiotic – it simply points to the stability of correlations between social status and physical and mental flourishing, without suggesting a mechanism for such a correlation – but it is implicitly biosemiotic: if the correlation exists (and Marmot's research says that it does), then, clearly, it is the result of the communication of signs. Marmot's research findings clearly refer to signs 'sent' and 'read' which are, literally, written in the bodies of those reading them in terms of health and mortality. In other words, one of the things which humans know tacitly (and experience unconsciously, but bodily, to a significant extent) concerns their place in human hierarchies; they have an understanding of the extent of (and the limits on) the opportunities they have to *affect* others. Thus, inasmuch as class, race, gender – and geographical place as well as social place – also describe people's ability to influence both what they can do (effect) and how others respond to them (who they can affect), these remain powerful determinants of healthy physical (and of course mental) flourishing.

What Marmot's work shows is that there is an absolute correlation between social 'place' – understood in Amartya Sen's sense as capacity for exercising a variety of 'capabilities'[16] – and health and life expectancy. As above, I would also include geographical 'place' as a central part of social 'place' – both within a country and between countries, especially in a globally communicative world. Above a certain level of basic need, the indicator of health and ability to flourish is not national wealth but relative equality and social integration. Those in a position to influence their environment, and in a position to deploy a wide array of capabilities and connections, have better health and live longer lives. The key to human flourishing is greater equality. Indeed, level of health, Marmot shows, is a good measure of the degree to which people are able to express their capabilities. The facts of discrimination and inequality, which increase people's sense of powerlessness, and exclude them from social and material resources – thus inhibiting their capacity for full expression of their capabilities – make people ill and chase their lives away.

Marmot's research demonstrates, in considerable detail, that human flourishing, creativity *and* productivity depend on the minimising of inequalities. In highly individualistic societies such as the UK and the USA, where utilitarian philosophies of possessive individualism have created steep hierarchies and low social integration and trust, deregulated marketisation quite simply leads to more ill health and more preventable early death. Marmot's research shows that '[s]ocieties that are characterised by social cohesion, whether rich like Japan, poor like Kerala, or somewhere in between like Costa Rica, have better health than others with the same wealth but less social cohesion'.[17] The pursuit of equality is not an empty idea, or idealistic denial of hierarchy and the significance of status; on the contrary, it is a recognition of the social importance, and effect, of these things upon human well-being, and an attempt to remedy it.

Philosophical rationalism, as developed in the philosophy of Socrates, Diogenes, and Empedocles, for example, was precisely the attempt to put forward a way of thinking which did not make human beings so vulnerable to the opinions of others, and to the need to defend their honour when it was challenged. As Schopenhauer put it much later (but in the same tradition), in *Parerga and Paralipomena* (1851), 'Other people's heads are too wretched a place for true happiness to have its seat'. In *Status Anxiety* Alain de Botton writes that it is because people's estimations of others are so often inaccurate, and because our emotional responses to things such as these so often overshoot the mark, that the Greek philosophers advised that all estimations be brought to the court of reason.[18] However, while thinking carefully about things (being reasonable), and not acting immediately upon one's impassioned emotional responses, is generally a good idea, over-rationalising in ways that *ignore* your emotional responses is very bad for you – so bad, indeed, that it makes you ill. This kind of rationalising repression, in which people accept as normal things (such as work overload in a macho office culture, for example) which really are not, makes people more susceptible to infections in the short term and also more susceptible to autoimmune diseases such as cancer in the long term.

Both Marmot and Putnam note that it is not your workload (whether at work or at home) as such which matters, but the amount of control you are able to exert in any particular situation. We might also note in passing that the possibility of rule bending or 'frame' stretching and changing is an important aspect of creative behaviour. Where this kind of semiotic freedom is limited by steep hierarchies of 'command and control', human creativity is equally limited. Marmot notes, for example, that Japanese car companies have a tradition of shallow hierarchies in which the contribution of ideas from workers at all levels is encouraged, and where bosses wear the same white overalls, and eat in the same canteens, as workers. Wherever they are located in the world, and thus whatever the nationality of the workforce, Japanese companies with Japanese management have higher levels of productivity and lower rates of product faultiness:

> Workers in Japanese auto plants had far more hours of training than in European or American plants. There was more job rotation in Japanese firms, so workers spent less time doing the same boring jobs.
> ...
> We could think of this as a virtuous circle. The iron-rod school of management says: be soft to workers and accept low productivity or crack down hard and force them to give of their best. (Some disgruntled worker put up a poster in one workplace which read: 'The beatings will not stop until morale improves'.) By contrast, the virtuous circle says that there need be no trade-off: health and autonomy of workers and health and economic success of companies are likely to go together, not be in opposition.[19]

The problem is, and as I have argued throughout this book, that we are *social* creatures, and the *individual* power of reason (of which we, who are the inheritors of the philosophical tradition of rationalism, have made so much) is weak in the face of the judgements of the group. We may know very well that 'the views of the majority of the population on the majority of subjects are permeated with extraordinary confusion and error',[20] and that our status is not a reflection of our worth as humans, but as the epidemiological studies cited above have shown, our knowing it may not prove strong enough to prevent it from blighting our lives.

These arguments about organisms as existing in an environment which is their generative field (in which they are capable of self-repair and reproduction), with the basic organism/environment coupling being at the core of every stage of evolutionary emergence and increased complexity, also hold true for biosemiotic human beings. Our primal relation to our environment remains within us, and its history is inscribed in environmental preferences – which once were tied to our survival and now are artificially managed. For example, just as our closest primate cousins do, we still like to retreat upwards to nests to sleep in when possible. But society and culture is also evolutionary; it, too, is our environment in which, to flourish, we must be capable of a full expression of our capabilities. This is not a matter of individual capability. Marmot cites repeated studies that show that when high status monkeys (and in other tests other animals, such as rats) are moved into a new group with other high status monkeys, so that a new hierarchy must form, formerly healthy and flourishing animals that find themselves lower down in the newly organised group, begin to exhibit poorer health and limited life expectancy. In the social environment that is an emergent feature of human creatures in a natural environment, health (etymologically related to both wholeness and wealth) is generally the most significant indicator of creatures able to deploy a rich variety of resources.

Every organism has its generative field in which it has an organistic capacity for self repair and creation. But if you change that field, you risk damage – 'put Acetabularia into sea water containing 1 mM calcium or a newt into a cold environment (less than 5°C) and neither will regenerate or reproduce'.[21] We understand perfectly easily that human beings have a natural generative field beyond which they will not prosper. What we have understood less clearly (although not incompletely) is that, for humans, the environment *really is* socio-cultural too. To the natural generative field, we must also add the idea of the biosemiotic sphere, or semiosphere. In the human environment, being richly connected in many relationships with one's fellows, being able to influence events around you, having an abundance of both natural *and* social resources in other words, creates flourishing emotional and physical health. But change that field – create steep hierarchies in which many feel profoundly limited and disempowered – and such human beings will not flourish, will become ill, and will die earlier.

Another way of talking about Marmot's findings is to say that this kind of epidemiological research uncovers the extent to which our experience of our generative field (both natural and social) is written in our bodies. We understand that our physical range (water, food, temperature) has natural limits beyond which we will cease to flourish and will, eventually, die; but our understanding of our social limits – the generative field of human sociality – remains poorly, or at least not widely, understood. In fact, the findings of the Acheson Report to the UK Government in 1998 (Marmot was a member of the Scientific Advisory Group chaired by former Chief Medical Officer Sir Donald Acheson) were well-received, and to some extent were acted upon (if quietly) by the Labour Government.[22]

Acheson's 1998 *Independent Inquiry into Inequalities in Health* took the view that human health should be seen as an index of general flourishing of capabilities, and that stress should be understood not simply in terms of 'too much to do' but, more precisely, as high demand 'in the presence of low control, and imbalance between efforts and rewards'.[23] The *Inquiry* adopted a broad definition of the elements that constitute human health and flourishing. Its account included within its remit: poverty, income, tax and benefits; education; employment; housing and environment; mobility, transport and pollution; nutrition and the Common Agricultural Policy; mothers, children and families; young people and adults of working age; older people; ethnicity; gender; and the National Health Service. But this view of the experience of human sociality as fundamental to a healthy and long life remains politically 'quiet' – no doubt because it challenges so thoroughly the still dominant ideology of individualism and market competition expressed in macho work-cultures.

Another way in which epidemiological research can be explored is through new understandings from within the new field of immunology called psychoneuroimmunology. And it is to these that I now turn.

PSYCHONEUROIMMUNOLOGY AND THE END OF MIND-BODY DUALISM: HOW 'OUTSIDE' GETS 'INSIDE'

Psychoneuroimmunology (PNI) has become, over the past ten to fifteen years, a well-defined field, with many articles and nearly 100 books published.[24] And this PNI research has fatally undermined the old idea that mind and body are separate, and made redundant the old question about how they communicate. As Paul Martin's psychoneuroimmunological arguments in *The Sickening Mind* demonstrate, emotional stresses of all sorts affect the immune system; they lower its resistance to infection by bacteria and viruses, and to autoimmune diseases such as diabetes, arthritis and cancer.[25] Starting with the skin, which is a part of the immune system and its central task of identifying 'me' from 'not-me', the immune system is one of the body's three great integrated systems through which the body and mind are interwoven. The traditional view has been that the immune system is an autonomous system which works largely independently of mind and behaviour; but, writes Martin:

> this view is now known to be fundamentally wrong. The body's three main regulatory systems – the central nervous system (which includes the brain), the endocrine system (which produces hormones) and the immune system – do not work in isolation from each other. On the contrary, they are intimately connected and interact with each other in many important ways. Events occurring in the brain can produce changes within the endocrine and immune systems through a variety of routes, including specialised nerve pathways and chemical messengers.[26]

Since the 1980s, psychoneuroimmunologists have made considerable advances in understanding these interactions, which are of two basic sorts: electrical pathways using nerve connections, and chemical pathways using hormones, neuropeptides and other chemical messenger modules. Candace Pert's groundbreaking work in identifying peptide receptors in the brain (hence *neuro*peptides) has been a very important part of showing how wholly integrated these systems are.[27] Martin writes:

> The specialized nature of these mechanisms strongly implies that they have evolved for a purpose – to enable the brain and the immune system to communicate with each other. One good reason for believing that the brain and the immune system are meant to communicate is that they are hard wired to each other by nerve connections ... The brain and the immune system speak the same language.[28]

Our brains appear to know what is going on in the immune system, and the traffic goes both ways.[29] Clearly, what we are dealing with here are biosemiotic complex systems (of 'communication' and 'speaking the same language'), involving both positive and negative feedback; and this is a good illustration of the way in which the body-mind-environment systems can be understood, in Hoffmeyer's sense, as a 'swarm of swarms'. I will return to this below.

As with molecular biologists' use of a language of 'encoding' and 'messengers' to describe the activity of DNA and mRNA (the 'm' stands for 'messenger') within the metabolism of the cell, discussions of the immune system and its relationship with the central nervous and endocrine systems are drenched in the language of semiosis. Pert writes how her own research 'used receptor theory to define a bodywide network of information ...[which was able] to provide a biochemical basis for the emotions',[30] and further notes that 'these receptors and their ligands [corresponding and interlocking neuropeptides] have come to be seen as "information molecules" – the basic units of a language used by cells throughout the organism to communicate across systems such as the endocrine, neurological, gastrointestinal, and even the immune system'.[31] The immune system is, Martin says, of a 'mind-boggling' complexity second only to the complexity of the brain itself.[32] Further, and as Pert's work shows, the fact that the limbic system, including the amygdala (sites of emotional activity in the brain), is strongly implicated in many of these exchanges is suggestive of the ways in which these messages are importantly messages about emotional states – feelings, and powerful intimations.[33]

As with Marmot's research, Martin also indicates that such emotional stresses are significantly linked to degrees of social isolation, and also to our experience of our ability to control the stressful situations in which we find ourselves: 'The extent to which we are disturbed by a stressor depends on several things, including our sense of personal control over circumstances and the support we receive from those around us'.[34] Unsurprisingly, experiments confirm that these are things similarly shared by other animals. 'It is hardly surprising that our minds – and those of other species – should be so attuned to a sense of personal control, since control over the immediate environment is vital for most organisms' survival. Control signifies autonomy, mastery and empowerment ... Lack of control means being a passive victim, carried along by the tide'.[35]

The perception of lack of control over your environment produces – in humans as well as other animals – a mental state known as 'learned helplessness', which bears many of the hallmarks of clinical depression in humans.[36] Again, in agreement with Marmot, Martin notes that these experimental findings about control 'also make sense of work-related stress ... The most stressful jobs are characterised by a combination of high demands and low levels of personal control – in other words, having little say in how or when to perform tasks'.[37] Perhaps it goes without saying that work under these conditions is a particular feature of labour in technological and bureaucratic modernity, where work is governed by systems imperatives, and E.P. Thompson's 'industrial time', rather than the 'cyclical time', or 'natural' time, of pre-modern, and contemporary agrarian, societies. When Jack Goody showed an African visitor around a factory in Cambridge where women were lined up assembling radio sets next to a clock where they 'punched in' their shifts, the visitor asked 'Are these slaves?'. He was, Goody reports, 'used to a much more individualised form of work organisation, where he himself decided when to go to and when to leave his farm'.[38]

As all the indicators strongly suggest, not only environmental control, but human sociality itself, is a significant aspect of healthiness. We do better (very literally, in terms of health, recovery from illness when ill, susceptibility to accidents and death) when we are richly socially integrated and supported. Communication, in other words, is a very important aspect of what makes for a mentally and physically flourishing life. Communication, or semiosis, goes all the way down – from our social relationships to the state of our immune system and our capacity both to resist and to recover from illness. Martin remarks:

> It is ironic to think that science has recently woken up to the importance of human social relationships for physical health, just as we are losing those relationships. In most industrialised nations the number of socially isolated individuals is rapidly increasing through the effects of social fragmentation, the breakdown of nuclear families and the growing number of old people. More people are living alone – some from choice, but many because they are separated, divorced, widowed or simply unwanted ... Current scientific knowledge suggests that we will pay a hefty price for the fragmentation of our society – in medical costs as well as human suffering.[39]

The epidemiological and psychoneuroimmunological work cited above is, unsurprisingly, supported by studies concerning the importance of social capital. Although the term was first invented in 1916 by L. Judson Hanifan, and disappeared and reappeared throughout the twentieth century, it was only in the closing decades of the latter that research into social capital developed into a distinct research field.[40] As Robert D. Putnam puts it: 'The idea at the core of the theory of social capital is extremely simple: Social networks matter. Networks have value, first of all, for the people who are in them'.[41] Health, well-being, and even happiness appear to be more closely tied to the extent of a person's social connectedness than to anything else:

> We describe social networks and the associated forms of reciprocity as social *capital*, because like physical and human capital (tools and training), social networks create value, both individual and collective, and because we can 'invest' in networking. Social networks are, however, not merely investment goods, for they often provide direct consumption value. In fact, the very large international literature on the correlates of happiness ('subjective well-being' is the accepted jargon) suggests that social capital may actually be more important to human well-being than material goods. Dozens of studies have shown that human happiness is much more closely predicted by access to social capital than by access to financial capital. In fact, the single most common finding from a half a century's research on the correlates of life satisfaction in countries around the globe is that happiness is best predicted by the breadth and depth of one's social connections.[42]

Of course, not all social networks are conducive to the greater good (Putnam cites the Klu Klux Klan). But an understanding of the relative goods (for those involved) of, say, street-gangs, or 'drug communities' – alongside their more socially acceptable forms in the shape of 'faith communities' or trades unions – can, at least, lead to a better appreciation of the social dynamics leading to their formation.

All humans understand (if not, necessarily, cerebrally) the importance of 'belonging'; the socially disenfranchised will tend to form networks on the basis of disenfranchisement itself. In these cases, the fact of social disenfranchisement will tend to be acted out in 'anti-social' sociality, since network membership is defined precisely in these terms. It follows from this that the appropriate response to anti-social behaviour is to understand it not in terms of anti-social individuals and their punishment, but in terms of the anti-social sociality of networked groups and the need for social inclusivity, and for programmes which pursue this. It also follows that the greater the social divisions between the socially 'included' and 'acceptable' and the socially 'excluded' and 'unacceptable', the greater will be the tendency for 'out' groups to form networks of sociality defined by acts which are 'anti', or hostile, to the networks perceived to have excluded them. This is how the human need for social esteem, and social capital, works.

How, then, can we get a clear theoretical grip on the data exposed by the research cited by Marmot, and Martin and his PNI colleagues, and supported by research such as Putnam's? Unsurprisingly, much of the detail uncovered by Marmot, Martin and Putnam has long formed an unproven part of human wisdom about well-being; but it is only in the past twenty to thirty years that the connections between social life, state of mind and health have been experimentally confirmed. What we need is a richer language for our descriptions; and, clearly, since we are wanting to be able to talk about the ways in which the social world (*Umwelt*) of human verbal language might also be related not only to the world of signification more broadly understood, but also to the biological language of the immune system and its conversations with the nervous system and the brain, and the endocrine system, what we are seeking is a theory capable of describing natural and cultural life alike, all the way up: from cell to society; from the inner environment of the organism to the outer environment in which it lives. This is what biosemiotics attempts to do. In fact, it seeks to provide a language for talking about the 'language' that not only permeates all life but, arguably, *is* the identifying feature of life itself.

A PERFUSION OF SIGNS: BIOSEMIOTICS

Biosemiotics developed from the confluence over time of three main sources. In a special 1984 issue of the journal *Semiotica* edited by Sebeok, the latter describes the development of biosemiotics as stemming from the semiotics of Charles Sanders Peirce; from the biology of Jakob von Uexküll as formulated in von Uexküll's *Bedeutungslehre* (*The Theory of Meaning*) (1940); and from his own development of zoosemiotics, the study of animals' use of semiosis.[43] What von Uexküll's work contributed was the idea of each organism having its own *Umwelt* shared by conspecifics, predators, prey and so on:

> The *Umwelt* is then the crucible of the animal's informational relationships, perforce the meaningful ones, with the environment. In fact, for the animal, the environment *consists of* these sign relationships, with conspecifics, with predators and prey, with shelter, weather and terrain, and with odors, sounds and silence. In many respects this anticipates and complements the notion of the *niche* as high-dimensional hyperspace.[44]

To this idea of the *Umwelt* (more or less identical to Goodwin's 'generative field' and Maturana & Varela's 'coupling') as the species environment in which things signify, or have meaning, von Uexküll added the idea of the individual organism's *Innenwelt* to which the *Umwelt* is the necessary correlation. As Sebeok writes: 'these correlatives mediate all constellations of sign traffic having relevance to the animal'.[45] Sebeok's work, drawing on Peirce and von Uexküll, showed

4. Perfused with Signs

that every organistic *Umwelt* and *Innenwelt* was, in Peirce's words 'perfused with signs':

> Until the recent decade, it was biologically acceptable to refer to two kingdoms of living things and to underline the unique properties of humans within the so-called animal kingdom. In this spirit, Sebeok in 1963 and later contributed 'zoosemiotics', distinguished 'anthroposemiotics', and anticipated 'phytosemiotics'. The last subject has been explicitly broached by Krampen (1981), and defended as a domain of study by Deely (1982).
>
> Although zoosemiotics and anthroposemiotics suggest some discontinuity between ethologically-grounded communication and species-specific genres of humans, this oversimplifies the more interesting realities, as has often been noted ... plants communicate complex messages, including tree-to-tree pheromonal warnings about caterpillar predators (reported by Orians and Rhoades 1970; Baldwin and Schultz 1983), and cell-to-cell memory within cotton seedlings providing increased resistance to mite predation (Karban and Carey 1984). This is but to indicate that study and documentation of plant-plant and plant-animal-plant communication has grown vastly over the past decade (Montalverne 1984).
>
> At still other levels, it is now recognized within biology that, rather than two, there are at least three (Woese 1981), if not five (Whittaker 1969; Margulis 1970, 1981), kingdoms of living things. Beyond the confines of traditional biology, moreover, our planet, Gaia, has evolved as a system profoundly shaped in many stages by its living constituents (Lovelock 1972, 1979; Bargatzky 1978).[46]

This work is explicitly developed within a systems theory, or complexity, perspective and, in the same *Semiotica* article, Sebeok wrote:

> It has often been noted that evolution in physical systems is associated with entropy – the decrease of order with movement toward equilibrium; this is in accordance with the second law of thermodynamics. Yet in living systems, evolution is associated with an increase in order; this complementary process has been called the second law of systems (Makridakis 1977), and is often referred to as negentropy. At the same time, nonequilibrium dynamics have become understood in physical systems largely through the contributions of Prigogine and co-workers, who also recognise ubiquity of nonequilibrium in living and cultural systems.
>
> Because of its recent evolutionary spurt – from cybernetics to dissipative structures – there is no monolithic 'systems theory'. This makes for an eclectic approach. It is less savoury as a technique, for some persons equate knowledge with control, not realising that, where semi-

otic components play a major role, the *more* we understand the complexities of a system, the *less* we should be confident of our power to manage it.⁴⁷

Hoffmeyer's argument in *Signs of Meaning in the Universe* proceeds on the basis of recognising a number of things which will become related in his account of biosemiosis. First, he recalls Gregory Bateson's observation of monkeys at play. What Bateson noticed was that the monkeys did not bite each other (they were playing not fighting), but that they signalled 'play fighting' by giving a sharp snap at the points where, in a real fight, there would have been a bite. Bateson concluded that the snaps were signs, and what they signified was a 'meta-message': 'This is not a bite'.⁴⁸ 'Not', Hoffmeyer goes on to observe, constitutes 'a rule as to how one performs an either-or operation',⁴⁹ and is the basis both of imagination (the capacity to think about something that is *not* there) and all subsequent human signification. The constitution of a boundary (is/is not) is, therefore, the basic prerequisite of semiosis. But the 'is/is not' abstract boundary which forms the basis of imaginative semiosis and language is physically present at the very beginning of biological life. The constitution of a boundary, a primitive membrane in which 'me'/'not-me' is constituted, is also the mark of the first form of primitive life: the cell. A cell (and, later, the symbiosis of cells into more complex cellular entities) thus inhabits, however simply, an environment (an *Umwelt*) which signifies: for example, here is food; here is not food. In the very simple *Umwelt* inhabited by the bacterium e.coli, for example, messages from the environment in the form of a sugar in a gradient cause e.coli's flagella to spin in a way that sends it towards the sugar. In the absence of this food, e.coli's flagella spin in the opposite direction so that it tumbles randomly. E.coli *recognises* the presence of sugar in a gradient and *responds* accordingly. Recognition and response, stimulus and life-enhancing reaction – however automatum-like at this chemical level of life – are primitive forms of semiosis.

In living nature, there is something like memory – something which does not exist in inanimate nature. Memory exists in an organism in terms of its capacities for regeneration: defence/repair and procreation. We might say, Hoffmeyer suggests, that 'the body remembers'.⁵⁰ The human immune system is a very good example of a swarm memory interacting (and interpenetrating) within our bodies with the other swarm memories of the nervous and endocrine systems, and with all the many 'not-me's which it encounters, recognises, and (usually) overcomes. The immune system (to simplify a 'mind-bogglingly' complex and clever thing) consists of non-specific defences (including our fundamental 'me/not me' membrane, the skin), and specific more complexly specialised defences.⁵¹ The specific immune defences depend on free-floating humoral immune functions (antibodies) and cell-mediated immune functions. Both of these are incredibly complex semiotic systems involving remembering (in the form of remembering the antigens of an earlier 'invasion'), copying antibodies (B-lymphocytes responding to antigens in a process known as proliferation) and, once in the blood-stream, antibodies latching on to antigens which are recognised and 'chomped up' by passing phagocyte white blood cells.[52] The other way organism's bodies are remembered is, of course, in the process of DNA encoding and recoding that goes on in procreation:

> Here we would appear to be faced with the most crucial difference between the living and the lifeless. All living systems are fragile and in principle everything is forgotten once it dies. Nevertheless, thanks to that ingenious process known as procreation, an inheritance is bequeathed to posterity. This inheritance is a very sophisticated phenomenon, the essence of which is seldom properly explained. The essence of procreation lies in a principle which we will call coding – or, even better, *semiosis*.[53]

If we accept the biosemiotic argument that life *is* semiosis – 'signs, not molecules, are the basic units in the study of life'[54] – we can begin to think about swarm intelligence (the evolution of complex structural interdependencies based on semiotic communication), and about 'swarms of swarms' (the evolution of even more complex structurally and biologically integrated complex systems), as Hoffmeyer suggests. And, looking at the characteristics of these systems, we may better be able to see the ways in which human self-consciousness and memory, in the development of language and culture, repeats, or reiterates, successful evolutionary systems that have already happened at a biologically prior stage – but does so at the (relatively) 'new' level of self-consciousness, language and culture. Hoffmeyer points to Peirce's contention that 'nature has a tendency to take habits', and says that we can indeed 'find some trace of such a "habituation" in the history of the cosmos:

> Peirce's theory – this tendency to take habits – appears to represent one of the poles in a continuous process of development, where the other pole could perhaps be termed 'anarchy'. Nature's tendency to reclaim its independence by means of new 'inventions'. For simplicity's sake let us call these two opposing forces in natural history *fate* and *freedom*'.[55]

Indeed, we can see this principle at work, too, in the most recent developments in genetic theory concerning the origin of life itself. In *The Century of the Gene*, Evelyn Fox Keller cites Freeman Dyson's suggested answer to the fraught (for biologists) question about the priority of what I have been calling (after Hoffmeyer) the two main biological forms of memory: 'Which came first – genes or cells? Replication or self-maintenance?':

In 1985 Freeman Dyson, another physicist, published a small book called *Origins of Life*. Revisiting Schroedinger's question ['What is life?'], he suggests that Schroedinger's approach exemplified a long-standing over-preoccupation with genes. Life, he argues, requires not just nucleic acid [DNA] but also a metabolic system for self-maintenance [the cell]; hence the overwhelming likelihood is that it had not one but two origins. The emergence of living systems as we know them could have come about as the result of the symbiotic fusion between two independently evolved prior subsystems – one a rapidly changing set of self-reproducing but error-prone nucleic acid molecules and the other a more conservative autocatalytic metabolic system specializing in self-maintenance.[56]

From the moment life begins with the production of a simple membrane instating the difference between 'me' and 'not-me', and the subsequent incorporation into 'me' – by symbiotic evolution, and the extension of 'my' membrane to complementary parts of 'not-me' (this question of the relationship to 'the other' obviously has a psycho-social, and philosophical, dimension also[57]) – we seem to be dealing with the evolution of biosemiotic systems (or systems of systems: swarms), with features (identity; 'mind'; memory; encoding/decoding; reproduction) that we can still recognise in ourselves, and in our behaviour. Complex animals are 'swarms of swarms' of such semiotic intelligence, some of which, in social creatures, go on to form external (to the organism), but *not separate*, swarms of intelligence (societies) themselves.

Individualism identifies an ontological difference between individuals and societies. But there is no such difference: culture *is* the new symbiosis of self-conscious organism and environment which describes *human* being: 'the consciousness is part of the reality, and the reality is part of the consciousness, in the whole process of our living organisation'.[58] The essential *continuity* of 'internal' world (*Innenwelt*) with 'external' world (*Umwelt*) is, however, something we will need to discuss further. The evolution of self-consciousness, language and culture is the emergent form in which human animals signify and remember; and through this they enter into a new natural stratum in which, for the first time on this planet, we can entertain the hope of bringing our intelligence to bear upon the future in more than the immediate ways characteristic of other animals. In the latter, later evolutionary forms exhibit more semiotic freedom than earlier forms. Human semiotic freedom, our capacity to bring our intelligence and other capabilities to bear on the future, is thus very closely tied to the human experience of hope. Indeed, hopelessness is a strong indicator of subsequent illness. One of the most significant collective ways we try to understand things better in order to intervene in our future is, of course, in the development of modern science; another way is in politics. A politics based on the desire to maximise human flourishing must be one based on maximising both equality and hope.

On the basis of emergence,[59] Hoffmeyer tells the story of evolution as 'nature's tendency to take habits', and as the evolution of ever greater levels of complexity and semiotic freedom through the interplay of freedoms and constraints, which he summarises thus:

> The earth's currents of matter and energy were gradually channeled through the increasingly complex recesses of all its many life-forms. Physico-chemical habits became biological habits. Primitive cells were organized into endosymbiotic patterns we call eukaryotic cells. Eukaryotic cells acquired the habit of working together as multi-cellular organisms which in the course of time adapted to the prevailing logic of the ecosystems. The stabilization of living conditions under this form of logic made both longevity and intelligence an advantage and, hence, the logic of the ecosystems was eventually shattered by the appearance on the scene of humanity with its formidable talent for bossing prehuman life around. But even human beings could not shrug off this knack of forming habits. Each civilisation is a manifestation of the way in which a new master plan is accepted, a plan that will significantly boost or diminish the unpredictability of human thought and deed.[60]

The advent of language established both a significant increase in semiotic freedom, and also the shared empathic *Umwelt* characteristic of human beings.[61] Languages, like human societies, are rule-governed. Human intelligence, like language, is, though, endlessly creative; and we might surmise that human intelligence, language, culture and society, too, are governed by the interplay of freedoms and constraints: absolute freedom, by necessity, is not possible; but too much rigidity is deathly. Rules and frames are necessary, but the possibility of the freedom to bend and break them in order to initiate new ones is central to human creativity. This will be discussed further in Chapter 5.

It should be possible to see, as Hoffmeyer argues, that, with the complex open adaptive non-linear system of evolution, we are dealing with, as it were, 'nested' semiotic systems, in which each progressive level incorporates, and is dependent upon, all earlier levels. The rationality of natural evolution consists of the same structure of 'from/to' incorporation and forward-directedness that Polanyi sees repeated in human skilfulness and knowledge. At each stage, from the cell on upwards, these systems are dual-coded, or what Hoffmeyer describes as both digital and analogical: the digital code (DNA) is what makes species reproduction possible; the analogue code produces the individual nature of each member. Both forms of code are essential. As Kalevi Kull argues, 'Signs means messages mean information. Biological information, however, is not a simple issue': organisms recognise and interact with each other as analogue codes in ecological spaces, but are carried forward between generations passively, as digital codes. Life (and 'self') does not exist until both the analogue and digital (cytoplasm and nucleic acid) are present. He argues that this principle of code-duality could be taken as a definition of life.[62] As with the synchronic and diachronic aspects of language and culture, code-duality – self-reference and other-reference – is everywhere, and, at every subsequent level, more semiotically free.

An environmental niche is always also a semiotic niche. Every environment is, at the same time, and necessarily, rich in 'information': sounds, odours, movements, colours, electric fields, waves of any kind, chemical signals, touch, etc. On this view, life is primally semiotic. And, in the light of complex systems theory, this should not surprise us. Just as the 'tree of life' which describes evolution itself is also found in other natural systems of bifurcation (plants; lungs; blood supplies; river deltas, etc), so we should not be surprised to find that that the elaborated system of semiosis discovered in human cultural evolution is also found, in simpler forms, at every stage of life. The organism-environment coupling is a form of conversation, and evolution itself a kind of narrative of conversational developments. We could even say, as others have, and Hoffmeyer does, that evolution can be understood as a kind of play in which the outcome can be described (as in all forms of play as we normally understand them) as education. Evolution (as we can see in the evolution of Nicaraguan Sign Language referred to in the Introduction) is the play and education of life forms, which lead to higher, emergent, levels of informational complexity.

The usual translation of *Umwelt* is 'environment', but in biosemiotic usage the word does not refer to the environment in general, but to the species environment as that is conveyed in the idea of 'my world', or 'the world as I experience it'. This is the socio-cultural, or biosemiotic, version of Brian Goodwin's 'generative field'. Needless to say, this includes 'the environment' as that is more usually understood, but it refers more specifically to 'my world' in the sense of my lived experience which will die with me. It is the world in which the family I am born into, the place where I live, the educational experiences available to me, the circle of my friends, and of my work, my gender, my class, and my race will all, according to the way my society is organised, either offer me opportunities for full expression of my capacities or, conversely, place limits upon them.

We can, then, rephrase in terms of semiotic freedom both Marmot's findings about the relationship between status, the free exercise of capabilities and health and flourishing, and psychoneuroimmunological research showing the relationship (or flow of messages) between *Umwelt* and *Innenwelt* in terms of vulnerability to both infection and autoimmune disease. Human beings receive messages from their environments which are also messages about how 'safe' they are in terms of their capacities for responding in life-saving (i.e. life-enhancing) ways;

these environments are both biological and socio-cultural; they consist of where you are (accommodation, work, neighbourhood, region, country, for example), how you get treated (family, education), and your material and social resources for responding. Flourishing, we might say, is about how much semiotic freedom you have.

CONCLUSION

What I hope my argument in this chapter has shown is that the state of philosophy and politics described by liberal individualism and its understanding of the relationship between organism and environment, or, more specifically in its own case, between individual, society and culture, is simply *wrong*. And it is so because that philosophy does not have (and, until very recently, could not have had) any way of understanding, via a theory of evolutionary stratification and emergence, how organic life, in the form of biosemiotic systems, necessarily developed individual organisms ('me'/'not-me') *as a part of* a complex biosphere and semiosphere environment based on the conjoining of what Hoffmeyer calls 'digital' (self-replicating) and 'analogue' (self-maintaining) systems. Earlier biological forms (the three integrated swarm intelligences of human biology, for example) are not superseded by the later ones (self-consciousness, language, culture), but continue to exist, alive and operative, within them as the substrate of evolutionarily later developments – as the relationship between human immune system behaviour and psychological experience of the environment, discussed above, indicates.

In developing the idea that minds and bodies, culture and nature, are quite separate things, alongside the related idea that human individuals are essentially monadic, self-interested, creatures rather than social through and through, the 'western metaphysics' which now threatens us with global cultural dominance has effected some very fundamental errors. These are reflected in the conditions of social alienation which are found, to greater and lesser degrees, in Western societies. Romanticism attempted a challenge to this view, but its 'unscientific' appeals led to its relegation to the (less important because not rationally 'hard-headed') sphere of the aesthetic. In the light of this, a biosemiotic politics would be one in which very many things would be done differently – not all of them necessarily very palatable to the sometimes overly rationalistic temper of modern liberalism. It is to a brief, and necessarily tentative, discussion of this, alongside a more detailed account of human creativity as the wellspring of all progress political and otherwise, that I turn in the chapter which follows.

NOTES
1. Hoffmeyer, *Signs of Meaning*, op. cit., p94.
2. Laughlin, *A Different Universe*, op. cit., p208.

3. H. Maturana & F. Varela, *The Tree of Knowledge: The Biological Roots of Human Understanding*, London: Shambala, 1998.
4. For an account of the ways in which the iteration of simple rules eventually produces complex patterns or structures, see J. Gribbin, *Deep Simplicity: Chaos, Complexity and the Emergence of Life*, London: Penguin, 2005, especially chapter 3.
5. Hoffmeyer, *Signs of Meaning*, op. cit., p113 *ff.*
6. Maturana & Varela, *The Tree of Knowledge*, op. cit., p100.
7. Williams, *The Long Revolution*, op. cit., p39.
8. T.A. Sebeok, 'Nonverbal Communication', in P. Cobley, ed., *The Routledge Companion to Semiotics and Linguistics*, London: Routledge, 2001, p14.
9. M. Anderson, J. Deely, M. Krampen, J. Ransdell, T. A. Sebeok, T. von Uexküll, *Semiotica*, vol. 52, 1-2, 1984.
10. K. Kull, 'Introduction: Entering a semiotic landscape', C. Emmeche, K. Kull, F. Stjernfelt, *Reading Hoffmeyer, rethinking biology*, Tartu: Tartu University Press, 2002, p7.
11. T.A. Sebeok, 'Pandora's Box: How and Why to Communicate 10,000 years into the Future', in M. Blonsky, *On Signs*, Oxford: Blackwell, 1985, p453. John Deely, however, argues that it is only at the close of the 4th century A.D. with Augustine of Hippo's use of the term *signum*, that we achieve the idea of the 'sign in general': 'The notion of sign in general was precisely *signum*, Augustine's Latin term proposed just as the 4th century closed to express the idea that the universe of human experience is perfused with signs, not only through our contact with the natural being of our physical surroundings in the signs of health and weather, but also through our contact with our conspecifics in discourse and trade, even in our contact with the divine through sacrament and scripture', 'The Impact of Semiotics on Philosophy', www.helsinki.fi/science/commens/papers/greenbook.pdf.
12. C. Taylor, *Sources of the Self*, Cambridge: CUP, 1989; A. MacIntyre, *After Virtue*, London: Duckworth, 1981.
13. See MacIntyre, *After Virtue*, op. cit., chapter 2.
14. R.D. Putnam, *Bowling Alone: The Collapse and Revival of American Community*, op. cit.
15. M. Marmot, *Status Syndrome*, London: Bloomsbury, 2004
16. A. Sen, *Inequality Reexamined*, Oxford: OUP, 1992.
17. Marmot, *Status Syndrome*, op. cit., p195.
18. A. de Botton, *Status Anxiety*, Harmondsworth: Penguin, 2005, p123.
19. Marmot, *Status Syndrome*, op. cit., pp131-2.
20. de Botton, *Status Anxiety*, op. cit., p125.
21. Goodwin, *How the Leopard Changed Its Spots*, op. cit., p162.
22. See Marmot, *Status Syndrome*, op. cit., pp263-4.
23. Marmot, *Status Syndrome*, op. cit., p.267ff; p.265.
24. See, for example, E.P. Sarafino, *Health Psychology: Biopsychosocial Interactions*, New York, Wiley, 1998; T. Cassidy, *Stress, Cognition and*

Health, London: Routledge, 1999; E. Bachen, S. Cohen & A.L. Marsland, 'Psychoimmunology', in A. Baum et al, *Cambridge Handbook of Psychology, Health and Medicine*, Cambridge: CUP, 1997, pp35-39; R. Ader & N. Cohen, 'Psychoneuroimmunology: conditioning and stress', *Annual Review of Psychology*, 44, 1993, pp53-85; S. Cohen & G.M. Williamson, 'Stress and Infectious Disease in Humans', *Psychological Bulletin*, 109, pp5-24.

25. P. Martin, *The Sickening Mind: Brain, Behaviour, Immunity and Disease*, London, HarperCollins, 1997.
26. Martin, *Sickening Mind*, ibid., p75.
27. Pert, *Molecules of Emotion*, op. cit.
28. Martin, *Sickening Mind*, op. cit., pp76-7.
29. Martin, *Sickening Mind*, ibid., p80.
30. Pert, *Molecules of Emotion*, op. cit., p27.
31. Pert, *Molecules of Emotion*, ibid..
32. Martin, *Sickening Mind*, op. cit., p67.
33. Martin, *Sickening Mind*, ibid., p79.
34. Martin, *Sickening Mind*, ibid., p121.
35. Martin, *Sickening Mind*, ibid., p142.
36. Martin, *Sickening Mind*, ibid., p144.
37. Martin, *Sickening Mind*, ibid., p145.
38. J. Goody, *Capitalism and Modernity*, op. cit., p2.
39. Martin, *Sickening Mind*, op. cit., p172.
40. R.D. Putnam, 'Introduction', *Democracies in Flux: The Evolution of Social Capital in Contemporary Society*, Oxford: OUP, 2002.
41. Putnam, *Democracies in Flux*, ibid., p6.
42. Putnam, *Democracies in Flux*, ibid., p8.
43. *Semiotica*, 1984, vol. 52-1/2
44. *Semiotica*, ibid., p12.
45. *Semiotica*, ibid., p11.
46. *Semiotica*, ibid., pp9-10.
47. *Semiotica*, ibid., p20.
48. Hoffmeyer, *Signs of Meaning*, op. cit., p6.
49. Hoffmeyer, *Signs of Meaning*, ibid., p9.
50. Hoffmeyer, *Signs of Meaning*, ibid., p11.
51. Martin, *Sickening Mind*, op. cit., p68.
52. Martin, *Sickening Mind*, ibid., p66*ff*.
53. Hoffmeyer, *Signs of Meaning*, op. cit., p13.
54. K. Kull, 'A Semiotic Building: 13 Theses', C. Emmeche, K. Kull & F. Stjernfelt, *Reading Hoffmeyer, rethinking biology*, Tartu: Tartu University Press, 2002, p14.
55. Hoffmeyer, *Signs of Meaning*, op. cit., p27.
56. Keller, *The Century of the Gene*, op. cit., p42.
57. This psycho-social aspect of incorporation of complementary 'not-me' into 'me' will be discussed in chapter 5 of the present work.
58. Williams, *The Long Revolution*, op. cit., p39.

59. Hoffmeyer, *Signs of Meaning*, op. cit., p36.
60. Hoffmeyer, *Signs of Meaning*, ibid., pp35-6.
61. Hoffmeyer, *Signs of Meaning*, ibid., pp34-5.
62. Kull, 'A Semiotic Building', op. cit., p15.

CHAPTER 5

The importance of creativity

Raymond Williams starts *The Long Revolution* with the chapter entitled 'The Creative Mind' because he wants to affirm that creativity is central to the process of the whole complex organisation of life, and that each and every human is a creative part of the complex evolving whole of human societies – which we discuss under the term 'culture'. The key repeated terms in his chapter are 'complex' and 'communication'. Art is simply a special category of human communication, and making and remaking in general, in which a culture most visibly thinks its way into what Polanyi elsewhere describes as the 'foreknowledge of a host of yet hidden implications which ... discovery will reveal in later days to other eyes'. Williams writes:

> Art is ratified, in the end, by the act of creativity in all our living. Everything we see and do, the whole structure of our relationships and institutions, depends, finally, on an effort of learning, description and communication. We create our human world as we have thought of art being created. Art is a major means of precisely this creation. Thus the distinction of art from ordinary living, and the dismissal of art as unpractical or secondary (a 'leisure-time activity'), are alternative formulations of the same error. If all reality must be learned by the effort to describe successfully, we cannot isolate 'reality' and set art in opposition to it, for dignity or indignity. If all activity depends on responses learned by the sharing of descriptions, we cannot set 'art' on one side of a line and 'work' on the other; we cannot submit to be divided into 'Aesthetic Man' and 'Economic Man'.[1]

What Williams wished to do was to formulate some kind of workable theory of what society is, and the nature of culture within it, in order to think more deeply about the long revolution which modernity inaugurates, and to discern what, in its revolutionary nature, remains nonetheless an experienced problem. His method was to ask some fundamental questions about society, culture, and the relationship between the individual and society, and then to trace the emergence of various cultural forms in the modern period. His concluding chapter, 'Britain in the 1960s' notes the contradiction between the abundance of an expanding consumer society and the continuing widespread sense of social unease. The repeated refrain in that chapter is that, despite the obvious gains of democracy and a sense of increasing individual autonomy, there remains a deeply unsettling absence of 'a realistic sense of community'; a 'crippling' 'lack of an adequate sense of society'; a disturbing capitalist version of society only in terms of the market and profit; and a decline of socialism as a meaningful alternative. Williams also notes there the growth in modernity of those nihilists of revolution in whom real creativity deforms itself into 'those parodies of revolution often achieved in modern history in the delinquent gang or even in fascism'.[2]

It is on the basis of this understanding of the importance of real creativity and communication as central and essential to human cultural evolution as 'a process of general growth of man as a kind' that I began this book with a discussion not of the development in modernity of cultural forms per se,[3] but with a discussion of the development of theories of complex human sociality as a part of cultural evolution. The loss of a sense of community (as well as the loss of skills and tacit knowledges[4]) is a constant theme in modernity (Marshall Berman is eloquent about this in *All That Is Solid Melts Into Air*[5]); but while actual communities can be oppressive as well as supportive, it seems to me that what is mourned as lost is simply the cultural acknowledgement of the fact of human sociality itself. But modernity is driven, above all, by scientific and technological development. A yearning for the lost organic community which attempts to float free of this, or an opposition of art to science, or culture to nature, or the placing of community in articulate language alone (Saussurean semiology) instead of in our enworlded body-minds and biology (Peircian semiotics; Sebeok and Hoffmeyer's biosemiotics), cannot have any real, significant or lasting, purchase in a world characterised by the emergence of scientific understanding. Thus, I have sought to frame my arguments about complexity theory and science by emphasising its lineage – not in cybernetic information theory or in the social physics of cellular automata (though both occupy overlapping fields), which are relatively uncontentious players in cultural theory – but in the science about which cultural studies is either uncannily silent or parodically deconstructive: evolutionary biology. With the introduction of biosemiotics into the discussion, I hope I have shown that biology is entirely a part of the field of semiosis which is properly the concern of cultural theory.

CREATIVITY AS BIOSEMIOTIC AND SEMIOSYMBIOGENETIC FREEDOM

I wish to conclude with a discussion of the importance of creativity in complex systems as the fullest expression of human communication, or

5. The Importance of Creativity

what Jesper Hoffmeyer calls 'semiotic freedom'. Hoffmeyer's own work draws in part on Lynn Margulis's argument that biological evolution proceeds not mainly via random genetic mutation and natural selection but, more importantly and more fundamentally, via the symbiogenetic mutation and natural selection of cellular life.[6] On its own, the genetic information encoded in DNA is inert, and does nothing. It is only in the metabolism of the whole cell that life – as complex cellular combination of DNA, organelles and mitochondria – is able to re-write and evolve itself in ever more complex forms and swarms. The evolution of life is not dependent simply upon genes encoded in DNA; it is dependent upon the creative symbiosis of (initially microbial) separate organisms. On this account, evolved life has the encounter between similarity and difference, the self-identical and the other, written into it from the beginning. On this increasingly widely accepted view of evolutionary biology, the motor of evolution is, thus, the encounter of identity with an otherness which is, nonetheless, sufficiently semiotically recognisable to allow of a productive encounter and negotiation, expanding a semiotic *Umwelt*, out of which new strata of complex life can emerge.

Microbial in its beginnings of life on earth with the symbiosis of a prokaryotic with a eukaryotic cell, symbiogenesis implies both biosemiotic recognition and creative translation, and the possibility, thereby, of the further symbiotic evolution of interpretative communities of swarm intelligences. All these on-going processes (from microbe to mankind) of the difficult but creative encounter between similarity and difference also affirm Peirce's observation of 'nature's tendency to take habits'.[7] The patterns in which life began, and habitually repeats, continue still in us. As Margulis argues – and as psychoneuroimmunological research confirms – we are, ourselves, made of such things. This should be less surprising than, perhaps, it is. When we say there is nothing new under the sun, we are referring to process (which is habitual). But process open to the other (which we also call creativity) is the signature of life, evolution and change.

This evolutionary insistence on the meeting of self and other is the case not only in evolutionary development of the biological lives of our bodies, but is repeated in the lives of our minds and our cultures too. As every commentator on creativity notes, the latter seems to come to us by way of a strange combination of passive alertness to hints and active keenness to interpret or translate them. This appears to involve the structuration which Polanyi describes – in the internalisation of (semiotic) hints in order to make something new – as 'disattending from in order to attend to'. The process, note, is of disattending from what is internalised (or incorporated, or made a part of the self) in order to attend to *what can be recognised as significant*: the microbial habit of semiosymbiosis remains. Similarly, in his meditation on the importance of aesthetic creativity as that which comes to us lively from the other, Derek Attridge has described this as 'the coming into being of the new in a remarkable openness of the mind to what it has not yet grasped', and as what Derrida once described as 'receiving it as a stroke of luck'.[8] When we are being creative, whether in the arts or the sciences, what we seem to be alert to are the message-rich intelligences of our *Umwelt* which are participative, and which spring from a deep immersion of self in the otherness of our world. Abstract propositional thought, with luck, can sometimes catch this. Creativity seems to be symbiogenesis in culture: not something at all new in nature, but the way that biosemiotic life, semiosymbiogenetically, makes newness in us.

It often seems strange to me that we should think of human biology, and the human world of society and culture, as a radical departure from the rest of the natural world. Biosemiotics assures us that it is not so. Creativity in culture and language reiterates creativity in nature; it springs, in metaphor as in symbiogenesis, from the joining together of different things in newly productive and complex ways. Biosemiotics thus recasts symbiogenesis in the form which I have coined as semiosymbiogenesis. But the translation involved in semiosymbiogenesis is never a reduction of the other to the same; it is a process of intersubjective world-changing and making. This is clear in Margulis's and Hoffmeyer's accounts of biological symbiotic evolution and emergence; but we can also see the same insight at work in Attridge's description of its human and cultural dimensions: 'If I succeed in responding adequately to the otherness and singularity of the other, it is the other *in its relating to me* – always in a specific time and place – to which I am responding, in creatively changing myself and perhaps a little of the world as well.'[9]

What began biologically as, perhaps, a failed microbial act of eating an other,[10] evolved eventually, in us, as both sensuous aesthetic hunger and ethical constraint together: each implicated in the other, as desire and its necessary limits, in regard to the continuation of creative life. One imagines the millennia of failure and success in which these co-operative tensions were brightly honed by natural selection. As Attridge (attending to the more recent human aspect – but the wider biosemiotic evolutionary implications remain) writes:

> This process of responding to the other person through openness to change is not dissimilar, then, to the one that occurs when a writer refashions norms of thought to realize a new possibility in a poem or an argument. As has often been remarked, the sense the writer has when this happens is that of achieving what one was seeking (finding the appropriate word in a poetic line; articulating the next stage of an argument) and would be accurately expressed not by 'At last, I have made something new!' but rather by 'At last I have got it right!' or even 'At last I have got it!'. Granted, it is presumably part of the writer's general

5. The Importance of Creativity

intention to compose sentences that are different from anything written before, and that at the same time are intelligible, informative, pleasure-giving. But what is foremost in the creative mind is neither the issue of innovation nor that of communication; it is the demand being made for a just and generous response to thoughts that have not yet even been formulated as thoughts, feelings that as yet have no objective correlative. In responding to the other person, to the other as a person, a similar demand for justice is at work, requiring a similar step into the unknown.[11]

But this step into the unknown is, perhaps, not without structure. Our condition, when presented with a good work of art, precisely resembles both that microbial appetitiveness and negotiation of appetite described in Margulis's account of symbiosis, and also the condition of the scientist described by Michael Polanyi as an intuition of 'as yet undiscovered things'.[12] For Polanyi, this is the structure of all complex development and 'tacit knowledge' in which we 'disattend from' (what has been internalised) in order to 'attend to' what comes to us as new and strange. Polanyi's 'intuition of as yet undiscovered things' is the condition of openness we experience, at our best, before the signs the artist offers us. And this is also our attitude in all creative thought itself. For the attitude before a work which we contemplate – in which we 'disattend from' everything we already know in order to 'attend to' the ways in which that work can inform us, until we have 'made it creatively our own', and lived in it and off it, but without exhausting it – is precisely the structure of all education.

When Buckminster Fuller said 'I seem to be a verb', he was pointing to the processual nature of being. We, like the rest of nature, are not transcendental a priori essences, but are open processes always opening upon what the world affords us. Our experience of what we call the object world, which for the past three hundred years we have been taught foolishly to think of as the experiencing of mastery, is in fact an experience of affordance. As Tom Stafford and Matt Webb say, 'objects ask to be used', and it seems that our relation to the object world is that of learned tacit knowledge: it is participatory, and, in it, the world which comes to us also, in the same moment, shapes our responding.[13] We humans also exist as co-dependent and co-arising in nature and in culture. Our ecology and creative adaptation thus includes both our natural and our cultural environments. The 'objects' that we meet, and which afford us possible actions, are poorly understood when only understood as separate 'objects' along the lines of the old Cartesian divide. What the 'object' affords us, if our life is not to be a series of deaths, is, as biosemiotics suggests, the responsibility of responsiveness. But this responsiveness – which artists and writers have often expressed in terms of a feeling of responsibility towards their creativity, a kind of lived ethos – is a responsiveness to process which emerges in the ecological relation between self and other, or, more abstractly, between similarity and difference. This structure of attentive responsiveness to otherness – a fundamental creativity and openness to newness evident in all life – is also at the heart of what, for humans, constitutes an ethical relation.

Natural growth, whether evolutionary or in any organism, has the direction of time's arrow. It is (but in the sense where we imply no final causation drawing everything forward) teleological and self-organising (autopoietic). The simple mathematical rules of the Fibonacci series produce the beautiful complexity of snail shells, pine cones, seed arrangements and branching periods in various plants. But, as Victoria N. Alexander, drawing on Peirce's lecture/essay 'Design and Chance' (1883-84),[14] points out, chance also plays a part.[15] Alexander, thus, characterises creativity as arising from the interplay of necessary rules and the chance experience (often lived as serendipitous) of *other* similarities. Peirce's triadic formulation of signs as iconic, indexical and symbolic can thus arrive at the evolution of articulate language – as Terrence Deacon argues: iconic similarity gives rise to the identification of indices via observation of the (iconic) repetition of indexical signs (this smoke and fire index *is like* the ones I've seen before); indices give rise to symbolisation in the recognition that one thing can regularly stand for another.[16] Once the latter idea is achieved, purely conventional (i.e. symbolic) relationships between a signifier and an otherwise unrelated signified which it stands for become thinkable. And finally, such conventional relationships also make possible further chances of new and previously unnoticed iconic similarities and repetitions in difference, giving rise to further symbolic generativity.

To say this is, of course, simply to notice that this is how languages themselves evolve. Very often we talk about this in terms of the productivity of metaphor, and commentators on creativity usually notice that it involves a mixture of scripts and frames and also, crucially, frame-breaking. Thus, Alexander argues that literary creativity is an emergent feature of the conjunction of systemic directionality and the originality afforded by chance. Not only objects constitute affordances; very obviously language does too. Directionality is law-like (Peirce's 'nature's tendency to take habits'), but also capable of original change where some interpretant (in our case a receptive and original mind; but chance 'recognition' of similarity and forms of symbiosis is also obviously possible in nature[17]) is capable of responding to a similarity between different things. For Alexander, such 'efficacious coincidence' is not only the motor of scientific, artistic and literary creativeness, but is also 'teleology's proper object of … study'.[18] Finally, we might note that the mysterious quality of creativity – its spontaneous and seemingly inexplicable emergence – may well be due to the fact that it draws upon ontologically prior (and thus partially occluded and 'disattended to') tacit knowledges of

5. The Importance of Creativity

'earlier' iconic modes of signification – of similarity within difference – which are old in human semiosis, and indeed, as symbiogenesis, as ancient as life itself.

In this way we may be able to see not only the forward-directed nature of creative life, but also that what we value in art is simply the most visible form of an ingenuity which not only belongs to all human beings, and is semiotically evident in all life, but which is, importantly, also the means by which human societies and cultures both grow and, inescapably, grow art and ethics (and also anti-ethical aesthetics) together. To ignore this, or to allow real creativity (as opposed to its glistering parodies) to be stifled, is to embrace defeat and decline. Just as cynicism is the negative mimicry of realism, so parody and pastiche are the negative mimicry of creativity. If it is necessary to make this argument again, and to keep making it, it is because, as human societies grow into ever more complex forms, they also produce accretions of power – first in institutionalised religion, and then in institutionalised secular bureaucracies and other blocs – whose general effects of vested interests tend towards limitations (first mythic and then rationalised) upon the creativity which all humans are inclined to bring, if at all possible, to their expansive labour on, or in, the world of the human *Umwelt*.

Williams began *The Long Revolution* with a discussion of creativity because his nascent (although impressive and substantial) grasp of social complexity and cultural evolution led him to tackle the question of the ways in which this is worked out in the creative acts of individuals that arise from the odd fact of *collective* generational creativity found in changes in dominant 'structures of feeling' (which seem to come from nowhere), and which act like waves in a tide – emerging, then at full force crashing on the shore, and gradually receding; Williams described these in terms of residual, dominant and emergent ways of seeing the world, through which a society and its culture develops. This clearly echoes Polanyi's 'disattending from in order to attend to' idea of the structure of tacit knowledge, in the movement from what is internalised and past, through the present conjuncture (which is also always pregnant with the forward-directedness of nature), which leads towards intimations of things as yet unseen. Williams understood, that is to say, what James Robertson has called the 'pre-political' work of culture and cultural analysis and synthesis: 'creative thinking and dissemination of ideas about social transformation belong, not to the processes and activities of mainstream politics, but to pre-political processes and activities aimed at getting new ideas and new policies on to the mainstream political agenda'.[19]

As far as political understandings are concerned, the focus on human creativity – which is always an embodied, partly tacit, and *social* process – doesn't get us directly to mainstream political action or manifestos; it gets us to the distal *conditions* (the emergent structure of feeling) under which these conditions might be formulated and enacted. But it is precisely because these conditions will not emerge without Robertson's 'pre-political' work – which is actually a description of creativity itself: the transforming of ideas about a field, or fields, of knowledge – that a focus upon the nature of creativity, and a preservation of *its* conditions, is so important. However, when talking about understanding creativity, and especially in biosemiotic terms, I am also talking about the ways in which such an understanding might inform 'pre-political' work itself. Following the bursting of 'Continental Theory' upon the anglophone world from the late 1960s onwards, and the related development of cultural studies as a new discipline in the 1970s, a problem – in the form of a silence – in cultural studies, the humanities and the social sciences in general, has remained. The difficult thing for modernity is biology.

Whether in animistic early societies, in Greek philosophy, or in the world's great religions, the power of tradition and placating the gods have been the means by which the human animal's biology (often referred to as 'the passions') has been constrained. The growth of rationalism in the eighteenth century, and the faith then placed in human reason, meant that things could be contemplated and done – both good, as in the increase in forms of emancipation, and bad, as in the ways in which human beings were treated in technologically modernising societies – which were often seriously at odds with our bio-social selves and needs. Of course, and as Freud pointed out, there is always a tension (which no human being ever entirely overcomes[20]) between instinctual life and culture; the development of society and culture is the means by which this is adaptationally addressed in human evolution. Nonetheless – and while our natural sympathies (now more often referred to as empathy), and our understanding of these, continue to progress and to inform our *Umwelt* – bios remains inseparable from society and culture.

As I have indicated throughout, and strongly in Chapter 4, bios forms an inner limit within the culture which extends its outer limits. Human happiness and well-being is almost certainly proportionate to the extent to which forms of social and cultural organisation are able to accommodate our creaturely inclinations. These turn out to be not particularly well conceived in rationalistic terms. Thus creativity and semiotic freedom are very important to our flourishing as individuals; but studies in another aspect of complex systems – concerned with group behaviours and the iteration of simple rules leading to complex patterns of behaviour – indicate that rationalist assumptions are not only very often misguided, but may also be powerfully ideological forms of social control (see, for instance, Philip Ball's discussion on human experience and use of space in his *Critical Mass: How One Thing Leads to Another*).[21] Whatever is the case, the result – possibly difficult to articulate, and mainly felt at the level of lived experience – is human distress.

When human biology began to enter into the debates in the nineteenth century, it did so within the constraints of Victorian society and culture so that evolutionary theory was framed, as Marx noted, within ideas about human society as comprised of individuals who were essentially competitive, and engaged, in Herbert Spencer's infamous formulation, in 'the survival of the fittest'. In the twentieth century – and with the growing understanding of genetics within a scientific mode of understanding which remained dominated by reductionism – discussions of human biology applied to society took the form of a sociobiological (now renamed Evolutionary Psychology) reduction of all human behaviour to the gene. It has only been quite recently – in the last twenty or so years of the twentieth and the first years of the twenty-first centuries – that a developing understanding of complex systems, combined with an increasing understanding of the ways in which biological systems are complex systems of stratification and autopoietic emergence responding to negative and positive semiotic feedback, has begun to make it possible for us to think scientifically about cultural evolution and biological evolution as contiguous systems – with the former 'nested' in the latter in the same ways as evolutionarily earlier strata are nested within subsequent strata in the human brain. With the development of biosemiotics, in which our understanding of ourselves as avid senders and readers of a profusion of signs is able to throw light upon both epidemiological and psychoneuroimmunological research findings, we are in a position to think again about the biological *and* semiotic life of human beings in non-reductive ways. This should make us more literate about our literacy *and* our bio-literacy.

Creativity is in many ways a word for describing autopoiesis as biosemiotic life: all nature and culture is creative becoming and change. In human complex systems, creativity is semiotic liveliness: liveliness in language, and liveliness in the processes via which tacit knowledge can emerge in concepts which can be articulated or, rather more accurately, *are* articulated as the process of such an emergence. As Margaret Boden argues in *The Creative Mind*, these processes also involve the use of rule-governed 'scripts' and 'frames', and innovative creativity seems often to arise from the ability to place temporary limits on these rules, or to switch or substitute scripts and/or frames. Creativity's affiliation with play and games is evident here in the novel manipulation of codes and recombination of scripts and frames. Again the parallel with DNA encoding and mRNA re-encoding and recombination should be evident: nature's 'tendency to take habits' goes all the way down and all the way up.

CREATIVITY AS CODE-DUALITY: TIME, THE INDIVIDUAL AND THE FUTURE

It is very widely accepted amongst psychologists that human creativity is combinational, that it lies, as exemplified in poetry for example, in the bringing together of different or disparate things in unexpected ways: a striking metaphor is the paradigmatic case. Boden gives the example of the poet Gerard Manley Hopkins's description of thrushes' eggs as 'little low heavens'.[22] The work of George Lakoff, Mark Johnson, and Gilles Fauconnier, indicating as it does the ways in which our language is dependent upon metaphors derived from our embodied enworlded experience, demonstrates how our cognitive processes, structured on the recognition of similarity and difference, burst into a new level of expressive creative freedom with the metaphoric extension that articulate language brings.[23] Such similarity and difference – the marking of 'me' and 'not-me' in the organism and its Umwelt – is characteristic, as we have seen, of all life. But to combinational creativity, Boden wishes to add also exploratory creativity and transformational creativity. What these latter two forms of creativity add to the first is – very importantly – the dimension of time. Exploratory creativity can only occur with deep immersion in a field (Boden thinks this typically takes about twelve years movement within any particular 'conceptual space'), and when a person has become so familiar with an area that they are able to 'tweak' it with confidence in order to explore the bending of its rules. Transformational creativity occurs when rule-bending gives way to an entirely new formulation of the dominant rules of the field.

Complex systems are characterised by a feature known as 'period-doubling'.[24] This is caused by the iterative nature of complex systems. At first, growth occurs in an orderly pattern of simple iteration, or simple doubling; but, as the pattern continues, in complex non-linear systems very small differences in the iterations begin to make themselves felt in the form of relatively sudden changes. At first these changes are random and chaotic, but eventually, over time, a new system of order – a new phase space – arises at larger, or more complex, scales. The system is more complex and more information rich. The limit of this behaviour in analogue biological life is the individual organism's generative field. At the digital level of DNA transmitted reproduction, the limit (in terms of populations) is any population's generative field. Period doubling appears to be a feature of evolution (and thus of culture and language) and, of course, it can be seen very clearly in the evolution of cultures where human creativity and communication is not overly inhibited either by isolation or by prohibitions in the form of tradition, religion or other ideologies.

Modernity, underpinned by the liberal philosophy of individual freedom, is thus revolutionary: the value it places on freedom and innovation (albeit mainly conceived of in crassly material economic terms) leads to speed of cultural change (as well as technological innovations in the speed of communication itself) that all commentators note. Modernity – which speeds up, and increasingly speedily joins up, the communicative world and, in so doing, also shrinks the experience of space – thus has postmodernity built into it. In his discussion of the latter, David Harvey has referred to this as the experience of 'time-space compression'.[25] Everything seems to speed up more and more, and, if we are not to be permanently over-stimulated and incapable of processing anything at all, we need to find ways of counteracting it. In particular, because creative freedom and innovation in modernity come to be conceived mainly in terms of economic freedom and market innovation – which are very partial ways of conceiving human creativity – we need a more comprehensive account of cultural creativity than that afforded solely by the idea of economic innovation, and individual capital accumulation. Creativity – the means by which human cultures evolve – is a *social* affair. Its contraction to the sphere of individual economic self-interest, especially as that is conceived of only in terms of the greedy present of an individual life-time, is quite simply an historical category mistake: creativity is not simply about either profit or the present. For the brilliance of human creativity lies also in its intimations of things as yet unseen which will be revealed 'in later days to other eyes'.[26] As with the evolution of sexual reproduction and mortality, human creativity – properly, if with difficulty, understood – is collectively, generously and responsibly, forward-directed: it embodies the inscrutable mystery (which Newtonian mechanics did not recognise) of time's arrow.

Thinking about creativity in cultural evolution, Boden's introduction of the dimension of time seems important. Where systems of positive and negative feedback in complex biological systems (including cultures) lead to iterations which produce period-doubling speed-up and chaos, the emergence of self-organisation (autopoiesis, or creativity) at new levels requires *time*. Postmodern time-space compression (the experience of cultural chaos which consists in a multiplicity of rapid and short-term 'innovations' in the absence of the 'evolutionary evaluation', tested by time, by which chaotic innovation is eventually subjected to the limitations inscribed in biological – or biosemiotic – life) needs to be countered by a mature understanding of, and investment in (in all senses), the ways in which reality (not merely a linguistic construct but a biological and biosemiotic constraint) fundamentally impinges on human animals and their world. Genuine innovative creativity, rather than simple novelty, in whatever sphere of culture it appears, needs thoughtful immersion over long periods of time. I shall come back to this near this chapter's end.

EVOLUTIONARY HUMAN CREATIVITY

Finally, then, and because the matter of creativity is fundamental both to autopoietic systems themselves, and also to thinking about them in pre-political terms, I want to turn to a brief consideration of some of the ideas about human creative life which I raised first in Chapter 3. We know that liveliness, or creativity, is a vital aspect of open systems. In biological systems creativity is at the level of *Umwelt* co-adaptation and organism-environment co-evolution. But in more complex animals, creativity (again, always a matter of creature and environment) takes the more complex form of inventiveness (not limited to primates by any means). In humans, the context of individual creativity is the evolution of the culture which is itself an accumulation of inventive responses to environmental pressures. Culture – meaning the state of inventive activity in a society at any one time – is the air we breathe, and it is upon this collectivity which individual acts of creativity depend.

A biosemiotic view of life is that all life, and all our experiences, are perfused with signs. But it is also clear that very many of these signs are tacit, either partially below, or well below, the level of consciousness (as in the relation between signs of status, of which we are only partially conscious, and the reading of these within our body of which we are wholly unconscious). In other words, I may feel angry or resentful about signs that I am unimportant to those around me, but I am quite unconscious of my rise in blood pressure, the depression of my immune system, and all the other physiological changes which make me, in the short term, more susceptible to infection, and, in the longer term, more susceptible to breakdowns in my auto-immune system. What I am interested in here is the way that these partially conscious, unconscious, and also perhaps once closer to consciousness but now forgotten, things are at work in human creativity and innovation.

Language, culture and society are emergent features of the social apes our ancestors were. Drawing on the work of Merlin Donald and others, Hoffmeyer argues (and this seems indisputable to me) that articulate language emerged from an earlier mimetic, i.e. embodied, language. We know, for example, that our predecessor species of 1.5 million years ago, *Homo erectus*, had culture, but without the cranial physiology which allows for the mouth and throat soft tissue which make articulate speech possible. Hoffmeyer speculates, further, that articulate language emerged from the various sounds and expressions accompanying mime and 'crystallized into standard sound patterns'.[27] As with the infant use of a 'sentence dictionary' in which one word stands for a whole sentence, we can quite easily see (and doubtless small children's ascent into language repeats this in significant ways) how more and more articulate sound patterns can be developed. Another model is provided in linguists' observations (discussed in the Introduction[28]) of the development of Nicaraguan Sign Language, in which what was very local and idiosyncratic, family-based, signing (idiolect) gradually evolved, with the greater contact with other deaf people which the Sandinista government organised in the form of national schools for the deaf, and noticeably with each new generation entering into these communities, into a highly articulated language – now internationally recognised.[29]

We can think about our creativity as born in mimesis, in gesture,

5. The Importance of Creativity

melody, wordless song, idiograms, representative drawings, and finally, articulate language. All these remain within us still as an emotionally powerful biosemiotic history. Our particular form of semiosis emerged, as the *Umwelt* particular to *Homo sapiens*, out of the narrative context in which an articulate sound pattern could stand for a sentence. As Hoffmeyer, quoting Bronowski, says: '"The normal unit of animal communication, even among primates, is a whole message"'.[30] The context in which this emerged must have been already powerfully mimetic and richly textured by sounds and made artefacts. All these are the basis of our continuing pleasure in narrative, in the song-like rhythms, beats and stresses of prosody and poetic language, in sculpture, painting and any surprisingly skilled and well-crafted thing. And, of course, in music. I surmise, as have others (Merlin Donald, for example, and, recently, Steven Mithen[31]), that the relationship between music and the gradual emergence of articulate language is old. Far from the popular image of early humans grunting, which they surely did, my guess is that much early human vocal expression – perhaps for the purposes of family-group calling, as in some species of gibbons – was closer to a kind of singing. Of all our mimetic capacities, music has by far the greatest power to move us with a terrific immediacy. That is why we use it to mark the emotional depth which we wish and expect to be attached to important ceremonial occasions. It also remains much easier to remember words (and numbers) if they are set to music; all of this suggests music's ancient provenance in the evolution of human articulate language. So I think we can offer the beginnings of a good semiotic account of our creativity in terms of the arts by reference to the semiotic origins of human culture. With the emergence of the *Homo sapiens* cranial physiology, neo-cortex and physical capacity for articulate language approximately 200,000 years ago, we step into that fantastically expanded biosemiotic field in which abstract thought becomes possible.

In fact, Hoffmeyer (following Sebeok) suggests that articulate language did not arise under the pressure of the need to communicate; some minimal form of effective communication existed for the *australopithecines* between 4 million and around 2 million years ago. This evolved into more effective forms roughly 2 million years ago with the emergence of *Homo habilis* and a correspondingly larger brain. *Habilis* skulls show evidence of the development of Broca's area in the brain which is associated with speech, or at least communicative skill. *Habilis* is the first tool user. Half a million years later, even more effective communication and brain size was successful for *erectus* for approximately 1.3 million years, until the emergence of *Homo sapiens archaic* between 200,000 to 400,000 years ago, then *Homo sapiens sapiens* and modern human brain size approximately 120,000 years ago. It's probably worth noting that the relative speeding up of brain size and communicative abilities which the evolution from *australopithecus* to *habilis*, then *habilis* to *erectus*, and finally *erectus* to *sapiens* describes repeats, on much longer time-scales, the generational speeding up of the development of Nicaraguan Sign Language that observers witnessed. This exponential speeding up is also a feature of language acquisition and use in children.

What Hoffmeyer conjectures to be responsible for the emergence of *Homo sapiens* and fully articulate language was not so much effective communication as such, but the capacity to think the abstract conceptualisation of 'not'. This abstract conceptualisation, he suggests, may be central to the move which defines the difference between mimetic *Homo erectus* and articulate *Homo sapiens*. Abstract thought, allowing the conceptualisation of a difference between 'me' and '*not*-me', may be the new membrane which divides *erectus* and *sapiens* and describes the conditions of emergence of the latter. Drawing on Gregory Bateson, Hoffmeyer suggests that 'not' is unrepresentable in mimetic culture, and that the capacity to think a thing which cannot be mimetically represented must mark the emergence of language. I think this is fundamentally correct. But I think that this capacity *must* be discoverable in late mimetic culture at least, and, indeed, that surely it can be found there. I would suggest, accordingly, that 'not' *can* be mimetically represented, and that this is done in the act of *erasing* a visual representation – a drawing in the dirt or sand, for example.

The development of iconic representation in drawing, I would suggest, is the most likely bridge – which also instantiates a gap and an emergence – between the culture of *Homo erectus* and the culture of *Homo sapiens archaic*. It is noticeable that a widely shared gesture for 'not', 'not that', or, as Hoffmeyer puts it, 'don't do that', is the waving from side to side of the hand and forearm (or its synecdoche in the wagging finger) which is identical to the gesture of erasure. Probably linked is the universal waving which signals the transition from 'I am here' to 'I am not here' and its converse 'I was not here, but now I am here'. The identification of 'not' as the abstraction which marks the absence of presence and the presence of absence is, thus, probably very closely tied to that special form of mimeticism in which a state of affairs is visually represented iconically, and doubtless sometimes also via the use of indexical signs. Drawing and painting, which also retain some considerable power to move us, are, for me, the best candidates for the conceptual 'not' which, as Hoffmeyer says:

> constitutes a puncturing of the space-time continuum which we innocently inhabit and take for granted, inasmuch as it presupposes an alienation, a non-participation – the essence of which is that one is neither that which is denied nor the denial. As such, the 'not' concept is – once it is no longer bound to a specifically negative action – no less

5. The Importance of Creativity

than the passport to the digital code, to language. And this key would appear to be contained – still unused – within the internal dynamic of the mimetic culture.³²

Merlin Donald, citing David McNeill's argument about the demonstrably close relationship between gesture and language, suggests that such gestures might, indeed, form the bridge between mimetic and linguistic culture.³³ One surmises that the absence of evidence for drawing and painting earlier than about 30,000 years ago is due to the very transience of the usual medium, dirt or sand, or charcoal or chalk on rock or wood, which was normally used. The use of the sort of dirt or sand available for permanent fixing when mixed with urine or eggs, for example, must have been discovered over time, and the skill required for the extraordinary accuracy and beauty of the representations of animals we see on the walls of the caves at Lascaux, for example – almost certainly made with a magical or sacred intent in those dark and inaccessible places – must also have developed over considerable periods of time.

With that move to the representation of 'not' in the possibility of erasure, and the conceptualisation of the negative, or negation as Hegel put it, we pass over that split, or gap, that Hoffmeyer writes of as inscribing both 'the analogic reality of experiences and the digital reality of language'³⁴ – from iconicity to indexicality to symbolisation – and also into a form of alienation. No wonder so many writers through the ages, not to mention the writer of the origin myth in Genesis, have depicted knowledge as the loss of innocence: the opening difference between experiential tacit knowledge and abstract conceptual knowledge. But this membrane between different kinds of knowing is, after all, just another membrane of difference, of the same order, although vastly different in magnitude and thus semiotic reach, to the one that evolved with the advent of cellular life. And, just as messages travelled between the cell and its *Umwelt*, so, similarly, messages continue to travel between the life of the body and our articulate *Umwelt* in culture – as epidemiology and psychoneuroimmunology now tell us. From that move across the divide between *erectus* and us, all human evolution will both contain all the strata which have gone before, and also continue in culture and language – spreading out in the greater *social* swarms of intelligence that we see in the evolution of ever more complex human societies. It almost seems as though the cell had globalisation immanent within it. Knowledge of this, what Nobel Prize-winning physicist Robert Laughlin, in *A Different Universe*, calls the move from the Age of Reduction to the Age of Emergence, may be the next membrane beyond which we struggle, as complex life has before, towards the larger symbiosis in which 'me' and 'not-me' can exist – with the recognition that subjectivity is not a wholly individual affair, but is *intersubjective* – within a larger intelligence of 'us'.

TIME AND OPENNESS

But, while I have offered a tentative biosemiotic account of the prehistory of our phenomenal creativity as a species, what might biosemiotics have to teach us about that creativity as it exists, now, in modern people in the present? That it, like the mimetic culture which gave birth to language, is social? Certainly: hence, for example, the phenomenon of simultaneous discovery noted by Margaret Boden in *The Creative Mind*. But what about our attachment to the idea of the creative individual: the genius? Well, let me close with some thoughts on what might constitute human creativity in general, and, more specifically, on those aspects of it which are probably more to the fore in the lives of those who we think of as particularly creative. I suggest that the latter are marked by a giving over of themselves – a willingness to surrender to – those parts of themselves, and us, which form the evolutionary substrate of abstract conceptual thought as it develops in articulate language and, eventually, in the linear rules of grammatical expression. In her discussion of associative semantic nets, Boden describes this as finding means of temporarily weakening some of the 'sensible cut-off rules' which normally govern the networks of associations through which our conventional 'scripts' and 'frames' for understanding things are organised.[35] Following Michael Polanyi's description of tacit knowledge, creativity, I will say, lies in being able to 'disattend from' the logic and rules of grammar, and other rule and grammar-like governed forms of semiosis, in order to 'attend to' the perfusion, and disorderly profusion, of many signs which aren't supposed to count as legitimate when semiosis is thought about only in linguistic rule-bound terms. This 'attending to' is both conscious and unconscious – a kind of free-floating attentiveness which Freud described as the necessary attitude of the analyst, and which poets have long described as waiting for the muse to descend.

In a letter to his friend Reynolds, the poet John Keats wrote that he had not been at work at his desk reading or writing, as he should, but that he had been in his garden, like a flower waiting for a bee, practising 'diligent indolence'. But note what Keats's metaphor also tells us. The flower which attracts the bee does so because the message of its colourful scented abundance is that it is nurtured by a rich environment. The bee itself, one of the social insects, is a longstanding metaphor of social, or swarm, intelligence. The individual plant is what it is because of the good things it has taken in from its own particular environment; its reproduction is fostered in the meeting of its located individualness with this bee part of the social hive.

We might say that being creative, then, consists, at least in part, of a rich, in some way well-nurtured, environment: a mind well-furnished with both experience and ideas – and this means richness provided by others: parents, teachers, friends, books and so on – and also a state of prepared receptivity. We might think of the former as a

5. The Importance of Creativity

cultural store, which is, in its narrowest sense, what we mean by education, and the latter – which is what education tries to inculcate in its wider sense – as a spirit of continual curiosity and attentiveness to what's going on – an openness to life. We can already see that we are speaking of semiotic freedom, here, and of its bequeathing across the generations. But this, alone, is still not quite enough. For one thing, creativity doesn't seem to come to us by consciousness and memory alone. Curiosity and openness means that we have to remain receptive; and very often this means that we seem to have to hang around, being diligently indolent, or reading books, or doing the washing up or going for a walk, or getting on a bus, or falling asleep, while some other process goes on of which we are quite unconscious. We see – with that strange forward directedness of life itself, and as Michael Polanyi describes scientific creativity in *The Tacit Dimension* and in *Personal Knowledge* – that there *is a problem to be solved*, and we think and think, but also we hover and wait. As Polanyi says, we have a general confidence.

But, also, and as Keats implies, we break some rules (I *should* have been at my desk, but I was in the garden), and this seems important. As articulate consciousness progresses in language and culture it becomes increasingly rule-bound as semiosis encounters grammar. But if we follow the linguistic rule that our thought proceeds according to an unfolding linear grammar, we will simply fall prey to the 'bewitchment' of language which Wittgenstein described when he said that 'Philosophy is a bewitchment of our intelligence by means of language'.[36] For creativity to occur, the linear grammatical logic of language abstracted from experience must give way to its earlier iconic substrates in the tacit experiential knowledge which lives in the body-mind. In order to be creative, we must, as Adam Phillips puts it, in a chapter called 'A Stab at Hinting' in the book where I first encountered Keats's letter to Reynolds,[37] be prepared to go along with the hints and hunches which are definitely outlawed as acceptable ways of going about things grammatically and in terms of merely conceptual knowledge. Poetry, of course, does this all the time. It listens (which is to say *waits* and *takes time*) for the happy co-incidence (chance) of rhythm and rhyme with a consequent renewal of meaning which is, thus, a kind of discovery. Poetry materialises and instantiates the process of human discovery in general. Poetry is the model of scientific discovery which, in *Personal Knowledge*, Polanyi describes as follows. I quoted this in Chapter 2, but it's probably worth repeating again in this context. Knowledge in science, he says:

> is not made but discovered, and as such it claims to establish contact with reality beyond the clues on which it relies. It commits us, passionately and far beyond our comprehension, to a vision of reality. Of this responsibility we cannot divest ourselves by setting up objective criteria of

verifiability – or falsifiability, or testability, or what you will. For we live in it as in the garment of our own skin. Like love, to which it is akin, this commitment is a 'shirt of flame', blazing with passion and, also like love, consumed by devotion to a universal demand. Such is the true sense of objectivity in science ... I called it the discovery of rationality in nature, a name which was meant to say that the kind of order which the discoverer claims to see in nature goes far beyond his understanding; so that his triumph lies precisely in his foreknowledge of a host of yet hidden implications which his discovery will reveal in later days to other eyes.[38]

So, what is the form our rule-breaking must take when we are creative? What do we actually *do* when we put on 'the shirt of flame'?

Let me tell a small story. It's a Sunday. I *should* have got up an hour ago. But Sunday's remaining associations of an allowable day of rest prevail with my fitful modern superego; so I linger in bed, listening to the radio at first, and then continuing, when the radio-alarm turns itself off, with my particular devotions (an obsession with a problem I'm trying to solve in the book I'm writing). I'm absorbed in a problem I've encountered, in reading epidemiologist Michael Marmot's *Status Syndrome*, about the vector by which a person's emotional experience of their life-world (another phrase for von Uexküll's *Umwelt*) can possibly be translated into real physical illness. I just can't figure this out. My youngest daughter, Tilly, appears and climbs into bed for a cuddle. We talk about various things for another hour or so; Tilly is wanting contact and affection, and I'm wanting to provide it. In the process of this idle engagement between mother and child, my eyes drift across the bookcase next to my bed and light upon a particular book: Paul Martin's *The Sickening Mind*. Something urges me to take the book off the shelf. I've bought it, sometime earlier, under some impulse I can no longer exactly recall, and have, as I see from various under-linings and margin notes, read some of it. I read some of it again and discover, in it, something which now begins to make a kind of sense to me that was previously unavailable. It tells me about psychoneuroimmunology, and reconnects me to a train of thought started seven years before when, researching an earlier book, I encountered the work of Candace Pert on neuropeptides in the context of Fritjof Capra's complexity book *The Web of Life*.

In other words, and by means I can't describe according to the linear logic of grammar or rule-bound behaviour, *The Sickening Mind* became a kind of key, like the falling into place of the tumblers listened to by an illegal lock-breaker, by which code I was able to read, and break into, what had been secreted away. But the bank I am robbing is the whole world of my unconscious tacit knowledge upon which, or, perhaps more accurately, in communion with which, the hard-won rationality of grammar will labour in order to write the rational linear sentences I will put into this book. This is a major conceptual achievement in my struggle with my problem; but it has arrived for me, by way of a number of illicit and un-rationalisable events and behaviours, in which I have given myself over to indirection, to affect and chance and human goods, and to the apparent serendipity of an unorganised Sunday morning. I don't think this experience will be at all strange to my reader; but how can we explain it?

What I have described above is just one event, but it isn't new or strange to me – any more than it is strange either to Adam Phillips or to the people whose particular engagements with listening for hints he discusses in the Chapter I mentioned above. Certainly this was, clearly, how Michael Polanyi experienced doing science. Candace Pert, too, in her book *Molecules of Emotion* writes of the passionate convictions which drove her research which she describes as:

> what I can only call a mystical process. It's like having God whisper in your ear, which is exactly what happened on Maui when I stood up with a slide of the HIV receptor in the brain and suggested a new therapy for AIDS, only to hear an inner voice say to me: *'You should do this!'*.
> It's this inner voice that we scientists must come to trust.[39]

This sort of science is what Brian Goodwin calls 'a science of qualities'. All this will make very difficult reading for people committed to our modern idea of rationality. But challenging the latter, and saying that it offers only a very limited account of reason, is what much of this book has been about. As I argued in Chapters 1 and 2, it is only when reason is taken at face-value as the bloodless, or dispassionate, operation of a mind divorced from an affective body, or, in other words, reduced to an ends/means calculation of special partisan interests, that atrocities seem rational and justifiable. Real rationality, the rationality of nature unfolding, is nothing like this.

When, many years ago, I first started following hunches as a researcher, I felt as though I was behaving somehow illegally. Surely, intellectual research is supposed to be an orderly and regular affair, perfectly lucid for the purposes of audit, targets and funding? And, of course, I'm not saying that one doesn't have to have a thought-out plan, and do the reading, and check what one is thinking by bringing it to the bar of reason in the form of sustainable argument, both with one's own critical voice and the voices of others, as a part of the process. But this isn't *simply* how discovery is. Of course, people can remain within the current paradigm of their discipline, and produce useful and publishable research; but original creativity involves risks and hunches which draw upon a form of rationality which still remains 'illegal', and not too much to be talked about, in our currently still plodding versions of reason.

Of course, *all* humans are creative, and my guess is that absolutely everyone quietly behaves in this 'illegal' way sometimes and to some extent – following hunches, even listening to dreams – in a process which is not rational according to currently acceptable accounts of reason. I doubt that there is a single person reading this book who doesn't recognise the mysterious process I am describing. It's only because our received idea of reason is that it is such a cool and dispassionate thing that, in fear of being not taken seriously in our work, we keep quiet about what we actually do, and thus contribute to a continuing – I would say ideological or metaphysical – conspiracy of silence. Of course, if, like Kekulé dreaming of the molecular structure of benzene, or Einstein reporting in a letter to Hadamard seeing Relativity in the form of pictures in his mind,[40] your strange insights prove experimentally, theoretically, or mathematically sustainable, the form of their origin in mysteries will be passed over, and the results accepted.

But what we're all *really* doing when we're trying to find out the 'more that is to be known' is being on the lookout for clues and hints – signs – which are available to us, in the first place, to the extent that we have done the reading, had a rich environmental earth and air and culture upon which to draw, *but also*, in the second place, to the extent to which we have been curious about, and open to, the world and to its rationality, the rationality of nature unfolding, in which we do have a kind of secret confidence – but about which we must not speak, at least if we are scientists, on pain of being labelled cranks. Because isn't it the case that, when writing a book, for example, or when writing anything that's supposed to present a good rational case, what you *plan* to do always has an uncertain existence which you *know*, if you have a little wisdom, hope and trust, will be fruitfully interfered with by other real things you know, but also know you don't know consciously, and which seem buried? It would probably be easier to write a poem. But the demands of *our* time, and they are good ones in their own way, mean that you must find ways of traversing again and again that gap between an experiential knowledge – which has now been internalised, and made unconscious and *tacit* – and conceptual knowledge. The *traverse* is the most difficult thing of all for us, because it involves a movement between what are, in our present culture, licit and illicit forms of knowing.

SWARM INTELLIGENCE
The body-brain-mind-*Umwelt*-biosphere-semiosphere continuum from which creative connections are made is, in the human individual, focused in associations (ideas, phrases, words, synonyms, metaphors, sounds, visualisations, sensations, including synaesthetic ones, present and remembered) which arise in the mind from millions of memories; these are marked by neural networks of associations, or patterns[41] – mainly in the brain, but presumably to some extent in the rest of the body too – or as memories of body-states, or the brain swarms' 'neuropeptide tones' as biochemist Michael Ruff puts it.[42] The extent and semiotic richness of

5. The Importance of Creativity 151

these connections can be judged by the fact that if you were to count each connection at a rate of one per second, your counting wouldn't be done until 32 million years had passed.[43] Connections are forged in experience. The human brain is quite plastic; every experience is registered, and strong connections (and memories) are forged by repetition while weak ones may fade. An infant learning to reach for an object is learning in this way from each swing of the arm (negative and positive feedback again), and the successful swipe in which chubby fist makes first connection with object – attached by some baby admixture of triumph and joy of consuming proportions, one imagines – is deeply marked, with all its neural experience of physical, spatial, muscular, balance, etc, co-ordinates as HURRAY-POSITIVE (the neural equivalent of the upwards 'air-punch').

We know that swarm intelligence is achieved by many units following relatively simple rules (making simple 'connections', or interpretations, and responses like E. Coli's flagella response to sugar in a gradient). Candace Pert's research on receptors, especially neuropeptide receptors, on the surface of cells with roots reaching deep into the cell's semiosis itself (whose presence can generate a number of varying responses from the cell itself depending on yet other 'messages' being conveyed at any one time in the system) also explained the swarm intelligence of the immune system and its intimate integration with the endocrine and nervous systems.[44] In *The Creative Mind*, where she discusses the possibility of genuine creativity in computers, Margaret Boden discusses the development of connectionist systems based on parallel-processing computers. These are not programmed 'but trained from experience by means of "self-organization"',[45] or autopoiesis. Boden likens one common kind of connectionist programme to a class of amazingly ignorant schoolchildren faced with the task of identifying whether or not the object on the teacher's desk is an apple. No child knows what an apple is, but each one knows both a small bit of information about apples, such as some apple colours, some small part of an apple's shape, and also whether her current opinion is consistent with that of her neighbour. Each child, Boden says:

> chatters non-stop about the tiny detail which is her all-consuming interest,[46] and her opinion is reinforced or inhibited by those of her neighbours who are themselves chattering to all their neighbours, and so on, so that the opinion of each is directly or indirectly affected by the opinion of every child in the class. The greater each child's confidence, the louder she shouts and the more notice is taken of her: Eventually, the children's opinions will be as consistent as possible (though there may still be some contradictory, low-confidence, details). At that point, the set of opinions considered as a whole is as stable as it is going to get: the classroom is in equilibrium.

The final decision is not made by any one child, for there is no class captain, sitting at a special desk and solemnly pronouncing 'apple'. It is made by the entire collectivity, being embodied as the overall pattern of mutually consistent mini-opinions held (with high confidence) within the classroom at equilibrium. The stabilized pattern of mini-opinions is broadly similar – though not identical – whenever the class is confronted with an apple (of whatever kind).[47]

This is an example of a Parallel Distributed Processing (PDP) system in which the parallel processing consists of *'localized computations'* with high connectivity.[48] It is recognisably a good example of a simple level of swarm intelligence characterised by a complex, although presumably not completely unchartable, multitude of positive and negative feedback systems. It is a complex system in which the determinations are, I assume, just about trackable. Compared to truly complex systems, or systems of systems, such as the human environment-mind-brain-body, though, in which determination and overdetermination produce a complexity of interactions which is deterministic in principle but not in fact, its complexity is very crude: child's-play one might say. It offers, relatively speaking, an example of a very small part of the complex patterns produced by the equivalent of neural nets governing the building of termite nests, for example. It is, needless, to say semiotic through and through. Its 'experience' is, initially, not, in fact, its own, but is fed into it by humans who devise mathematical ways of describing, in as much detail as possible, tiny aspects of 'appleness', and distributing these among the units making up the neural net. Nonetheless, PDP systems can and do learn.

But Boden's point is not to pit computers against humans, but rather to see if what computers can do is able to throw any light on how humans do what they do. Neural nets can give us a glimpse of a shadow of something like what goes on in complex human systems. They also reinforce some of the things that sane human beings understand about semiotic social systems. No man is an island. As in Marmot's epidemiological research, Putnam's sociological research, and the research of the psychoneuroimmunologists, high connectivity is important; it is a significant indicator, unsurprisingly, of the extent of semiotic freedom, which is, in turn, an indicator of good health. Being 'in the loop', being able to communicate and get heard, is important. Compare the prospects for semiotic freedom of the child of an economically anxious, relatively socially isolated, tired, worried or depressed, perhaps single, mother – for whom the communicational demands (love, attention, conversation) for interaction from her child may well seem impossible to meet, and who may thus sometimes react so as to close down these demands – with those of a child with an unworried, well-supported, mother with sufficient time and energy. Most people already *know* these things, of course, but it may be helpful to have a slightly different, and research-evidence-supported, nexus of languages in which to

5. The Importance of Creativity

describe them. Consider also the fate of the child whose mother, while relatively well materially supported, is depressed following her husband's infidelities (which we must also think about as communicative acts – and ones often very well read below the level of consciousness, if not consciously). She, too, may be made physically or mentally 'sick' by her semiotic experiences and, as a depressed or otherwise ill mother, be unable adequately to respond to her child's needs for the communication of love in all its many different forms. This, too, will be a relatively impoverished and damaged child.

Again, only psychoanalysis – largely through object relations and attachment theory – has had anything significant to say about these states of affairs that modern liberal societies have tended to view primarily (although not exclusively, there being recognised tensions) in terms of individual choices and freedoms. The ways in which we are semiotically linked, and our emotional and physical wellbeing is correspondingly linked, surely requires a much more widely understood sense of emotional literacy, and of the very real effects of its absence. Communication (in any organisation from family through company and institution to state and globe) is extremely important, both positively in terms of interconnected lateral webs or nets, and negatively, when these are limited by hierarchical cut-offs in which flows of information across levels are discouraged (as in command and control models). The gospel of liberal individualism needs to be constrained by our knowledge of the very real importance of human sociality.

Creativity, I will say, lies in the means of increased semiotic freedom. The development of conceptual language is a leap in the direction of greater semiotic freedom, and this latter provides the conditions for development and progress. We can think about this in terms of access to a wide range of symbolic languages. Creativity is impossible in the absence of structure, although it almost certainly requires forms of illicit behaviour such as rule bending or breaking. Languages are structures with different possibilities. Thus, for example, the arithmetical language afforded by Arabic numerals is much more productive than that offered by Latin numerals. The development of the concept 'nought' in Arabic mathematics allowed advances which simply weren't possible with the Latin form. Similarly, the language of Roman Catholicism – in which I'm including structures of power and interest enunciated therein in the role of the priesthood – was enormously rich in all sorts of ways, but sclerotic and over-constraining in relation to the development of the new language frames of science in Europe in the seventeenth century.

Semiotics is built into nature, and we are the animals in whom it has most richly flourished, and who have moved from what we call nature to what we call culture – though they are differences only of gradation in the direction of complexity and conceptual abstraction. Beneath this articulate conceptual knowledge, the language of the mimetic culture of

Homo erectus is still very alive. As Merlin Donald, citing Eibl-Eibesfelt's research, argues:

> [The] evidence demonstrates how the mimetic level of representation survives under the surface, in forms that remain universal, not necessarily because they are genetically programmed but because mimesis forms the core of an ancient root-culture that is distinctly human. No matter how evolved our oral-linguistic culture, and no matter how sophisticated the rich varieties of symbolic material surrounding us, mimetic scenarios still form the expressive heart of human social interchange.[49]

The richness of our languages is a direct result of our combination of curiosity and unparalleled adaptability in responding to various environmental pressures as we spread out across the globe, doubtless sometimes fighting those new groups we met, but also trading objects and ideas. Our semiotic richness consists in spatial, embodied, as well as narrative forms. Indeed, we can see our analogising body-grounded minds making the metaphors in which conceptual language is extended: a narrative is like a life; a life is like a journey (even Neanderthal grave goods tell us that they, like us, thought of death as a different kind of journey); a life is like a journey, and also like the progress of a day or a year.

The development of our language is like a journey also, and was doubtless facilitated by our journeying, by the bringing together of more groups into bigger groups in towns and seaports, with markets, and more and more exchange of signs as things and signs as ideas. With all this, our semiotic freedom and richness grew along with our creative capacity for innovation. With better stocked minds, we grew freer for invention. As Boden notes, people whose creativity is such that their talents or ideas have historical force are nearly always characterised by deep immersion in a field. They have what she calls 'a better sense of domain relevance' so that the chance of serendipity is more semiotically *meaningful* to them: 'These rare individuals, then, can search – and transform – high-level space much larger and more complex than those explored by other people. They are in a sense more free than us, for they can generate possibilities that we cannot imagine'.[50]

NEW FRAMES: POLITICS AND ETHICS

With the eventual frame-breaking of mythic thinking, modern science was able to institute a powerful new frame for what counts as knowledge, and also new scripts in terms of appropriate scientific procedure. This new level of semiotic activity has focused on observation, analysis and measurement. Its objectivity has consisted in the guarantee provided by collective reproduction, or repeatability, of experimental data, and, to a certain extent, on predictability. However, not all science

5. The Importance of Creativity

is concerned with predictability and mastery in this sense. Being able to describe the mechanisms by which something occurs in nature is, as in the case of Darwin's theory of evolution, not necessarily to be able to predict what will occur in the future. Darwin's theory was magnificently right in some important ways, but almost certainly incomplete – as recent developments in the genetics, and molecular and developmental biology, of complex systems suggest.

The development of our understanding of life in complex systems theory, of stratification, emergence and self-organisation – as in our improved understanding of creativity itself as emergent from semiotic complexity – does not dispense with analytic reduction, but adds upon it another evolved capacity in our cultural understanding. Evolution does not discard those successful developments which have gone before but enfolds all earlier developments within the latest, as we can see in the developmental architecture of the human brain. This, it seems to me, is what Laughlin's Age of Emergence will do too. Especially when viewed through the lens of biosemiotics – in which information and meaning are not seen as something entirely unprecedented, found only in humans, and categorically different from the rest of nature – this latest development in semiotic freedom is just the most recently arrived at stage of natural-cultural evolution: it is the stage in which energy *is* information, and information can, under the right conditions, become life in a universe which is, and perhaps always has been, 'perfused with signs'.

From a political view, there is much in what I have discussed in this book that will be congenial to socialists. The idea that flourishing – in Sen's terms the fullest expression of an individual's capabilities, or in Hoffmeyer's 'semiotic freedom' – is best brought about by greater equality, better education, richness of social contacts, social and economic empowerment, more loving kindness all round, will be welcomed. But an understanding of biology in culture – biosemiosis – does not conform entirely to the political lines and formations drawn up in modernity. Semiotic freedom, as attention to the correlative idea of the human *Umwelt* must make clear, is not just a jazzed up kind of liberalism, because it is not simply an *individual* thing – even though it is lived in individual lives. Semiosis is intersubjective and interactive; semiotic freedom is distributed freedom. It is not just my freedom to send whatever signs I want whilst remaining, at least consciously, an entirely selective reader of the signs I receive. It is a freedom constrained by the web of signs which signify in the human *Umwelt*. These are both natural and socio-cultural. Environmental damage, in this sense, means both damage in nature and damage in culture; these are not, essentially, different things. Environmental literacy must be understood to encompass natural, social, cultural and, by implication, emotional literacy also.

But the dimension of time which Boden's account of creativity introduces should alert us to another aspect of the ways in which a biosemiotic understanding might inform pre-political thought, though perhaps in ways which do not sit easily with the political affiliations of modernity. The wisdom which the unfolding of time (deep immersion in thought and practice in a field) allows is more commonly associated with a conservative frame of mind (a conservative politics and perhaps a politics of conservation) than with the radicalism of fundamentally 'new' or revolutionary ideas associated with liberalism and some forms of socialism. The short-termism which dogs capitalism's search for quick profits is radical in all the wrong ways: it is continually upsetting and destabilising of the longer time-scales involved in really creative exploratory and transformational innovation. A new radicalism concerned with cultural innovation and evolution might need to be a mature one in which the importance of time and human process was properly understood and accommodated. Needless to say, an entire society obliged to model its every action on the continually disruptive model of short-term market innovations cannot do this.

If the problem with modernity was, as Alisdair MacIntyre argues, that, in separating identity from social role or 'place', it made ethics or morality meaningless, biosemiotics puts human beings back in place.[51] But this is a place in a natural and cultural web of relatedness in which individual flourishing cannot be understood solely in terms of economic flourishing, and abstracted from emotional flourishing; and the individual, thus placed, cannot be abstracted from the flourishing of the whole. This implies some things which will be less congenial to liberal, including some socialist, ideas of freedom. An expanded idea of environmental literacy must mean that none of us are free to damage either the animate and inanimate nature around us, *or* nature in each other. Nature isn't just 'out there', but is *in us all*. This is a different way of thinking about humans as being 'in place', in which we discover ourselves as being in place, not only in virtue of social role, but in virtue of our being placed as processes of being in a processual web of natural, social and cultural life. This sense of place as mobile, processual, creative, and as inter-related and intersubjective, of course reintroduces an ethics. It is the semiotic ethic of responding: both responsibility and responsiveness.

In our common usage, to be a responsible person is often thought about in terms of being practically sensible, or realistic, in a way that generally excludes related terms such as 'sensitive'. In Western cultures still quietly dominated by the Cartesian division of mind from body, 'sense' and 'sensibility', 'sensible' and 'sensitive', still belong to different, divided, registers in which what is 'real', or 'realistic', doesn't include the very definite reality of feelings, ideas, intimations and other real, but not easily scientifically measurable, things. But 'responsibility', as the word suggests, actually refers us to semiosis and to an ethos of *responsiveness* in which *all* signs *matter*, i.e. are material and real, and are properly acknowledged as such – and are *read*. Acknowledging signs sent – whether from a culture familiar to us, from different cultural traditions, or from the world of animals and natural things more generally – is, of course, not straightforward – as anthropologists, ethnologists and ethologists have been obliged to acknowledge. Meaning is not transparent: we may misread cultural difference as sameness, and, where the natural world is concerned, we may anthropomorphicise. But no matter how heavily our culture lies upon us, we should not forget that we are animals too, amongst other animals, and are not entirely without the capacity for responding to the natural world in which we also have our archaic being. The difference is that, with us, biosemiotic responsiveness takes the form of an ethos.

What all *this* means is that the *semiological* (Saussurean) escape from nature, in which human meaning is believed to be restricted to articulate language alone, must give way to a wider *semiotic* (Peircian) understanding, in which embodied acts and deeds are more clearly understood as meaningful signs also. Thus, the liberal idea of human reason as that which can be rationally articulated in the abstractions of language must make room for Pascal's 'reasons of the heart'. Embodied and enworlded reason, thus expanded, takes account of environmental damage in a different way. In this understanding, the 'freedoms' of the liberal subject must be constrained by a newer kind of knowledge in which acts and deeds, not just words, signify. In this, your liberal freedom to abandon concern for others, for example – as though 'others' were not intimately related to you in the natural and socio-cultural web – must be set against the material and emotional environmental damage your irresponsibility and unresponsiveness will effect.

As a materialist philosophy and politics, socialism has always been interested in material conditions. To this, it can now add the recent scientific understanding, described in this book, that material life is always also affective life: what happens in people's emotional, psychological, lives *is* what happens in their material, physiological, lives and futures also. A progressive politics (and progressive policies) will be one which is able to make compelling scientific arguments for greater equality, for redistribution of wealth, for different work regimes, for better housing and environments, for local and global social justice, and for human flourishing understood as consisting in the good quality of all life. And while all this is unlikely to move those interested in preserving their own privileges in the short to medium term, over time the gradually accumulating weight of evidence will eventually do its work; and eventually this will, at the very least, serve to make the ideological nature of the resistance clearer.

The ideological universe we have inhabited for the past two to three hundred years is passing, and we enter, to use Robert Laughlin's title phrase, 'a different universe'.[52] This will mean the serious reconsideration of many liberal ideas. Human existential freedom in culture is expressively (semiotically) very great, but it is also limited by the biology upon which it is ontologically dependent. This imposes constraints upon freedom as liberally and rationalistically conceived; not because reason is the enemy but because our idea of rationality has been far too narrow. Our history is almost certainly the history of how we have managed, and will continue to manage, this dialectic between the imagination of reason and freedom, and our biological, and emotional, limits.

NOTES
1. Williams, *The Long Revolution*, op. cit., p54.
2. Williams, *The Long Revolution*, ibid., p382.
3. Williams, *The Long Revolution*, ibid., p59.
4. F.R. Leavis & D. Thompson, *Culture and Environment*, London: Chatto & Windus, 1960 [1933].
5. M. Berman, *All That Is Solid Melts Into Air: The Experience of Modernity*, London: Verso, 1983. See especially, chapter V: 'In the Forest of Symbols: Some Notes on Modernism in New York'.
6. L. Margulis, *The Symbiotic Planet: A New Look at Evolution*, London: Phoenix, 1999.
7. J. Hoffmeyer, 'Origin of Species by Natural Translation', www.molbio.ku.dk/MolBioPages/abk/PersonalPages/Jesper/Hoffmeyer.html, p6.
8. D. Attridge, *The Singularity of Literature*, London: Routledge, 2004, pp23-4.
9. D. Attridge, *The Singularity of Literature*, op. cit., p33.
10. L. Margulis, *Symbiotic Planet*, op. cit., p82.
11. D. Attridge, *The Singularity of Literature*, op. cit. p34.
12. M. Polanyi, *The Tacit Dimension*, London: Routledge & Kegan Paul, 1967, p24.
13. T. Stafford & M. Webb, 'Hack 67: Objects Ask to be Used', *Mind Hacks*, Sebastopol CA: O'Reilly Media, 2005, pp218-221.
14. C.S. Peirce, 'Design and Chance', *The Essential Peirce: Selected Philosophical Writings*, Vol. 1 (1867-1893), eds., N. Houser & C. Kloesel, Bloomington: Indiana University Press, 1992.
15. V.N. Alexander, 'Hopeful Monsters: Literary Teleology and Emergence', E:CO vol. 7, Nos. 3 and 4. (Forthcoming 2006)
16. T. Deacon, *The Symbolic Species: The Co-evolution of Language and the Brain*, New York: Norton, 1997. See especially, chapter 3, 'Symbols Aren't Simple'.
17. Margulis, *Symbiotic Planet*, op. cit.
18. Alexander, 'Hopeful Monsters', op. cit.
19. J. Robertson, *After The Welfare State*, 1980, www.jamesrobertson.com.
20. D.W. Winnicott, chapter 1:'Transitional Objects and Transitional Phenomena', *Playing and Reality*, London: Routledge, 1991.
21. P. Ball, *Critical Mass: How One Thing Leads to Another*, London: Arrow

Books, 2005, see especially chapter 6: 'The March of Reason: Chance and Necessity in Collective Motion'.
22. Boden, *Creative Mind*, op. cit., p41.
23. G. Lakoff, *Metaphors We Live By*, Chicago: Chicago University Press, 1980; Lakoff & Johnson, *Philosophy in the Flesh*, op. cit.; G. Lakoff & G. Fauconnier, *The Way We Think: Conceptual Blending and the Mind's Hidden Complexities*, New York: Basic Books, 2002.
24. J. Gribbin, *Deep Simplicity: Chaos, Complexity and The Emergence of Life*, London Penguin, 2005. See especially chapter 3: Chaos Out of Order'.
25. D. Harvey, *The Condition of Postmodernity: An Enquiry into the Origins of Cultural Change*, Oxford: Blackwell, 1989.
26. Polanyi, *Personal Knowledge*, op. cit., p64.
27. Hoffmeyer, *Signs of Meaning*, op. cit., p110.
28. Introduction to this work, p33.
29. A. Senghas and M. Coppola, 'Children creating language: How Nicaraguan Sign Language acquired a spatial grammar', *Psychological Science*, 12, 4: 323-328; R. J. Senghas, A. Senghas, and J. E. Pyers, 'The emergence of Nicaraguan Sign Language: Questions of development, acquisition, and evolution' in J. Langer, S. T. Parker, & C. Milbrath (Eds.), *Biology and Knowledge revisited: From neurogenesis to psychogenesis*, Mahwah, NJ: Lawrence Erlbaum Associates, 2004.
30. Hoffmeyer, *Signs of Meaning*, op. cit., p110.
31. M. Donald, *The Origins of the Modern Mind*, op. cit.; S. Mithen, *The Singing Neanderthals: The Origins of Music, Language, Mind and Body*, London: Weidenfeld & Nicolson, 2005.
32. Hoffmeyer, *Signs of Meaning*, ibid., p111.
33. Donald, *The Origins of the Modern Mind*, op. cit., p220.
34. Hoffmeyer, *Signs of Meaning*, op. cit., p111.
35. Boden, *The Creative Mind*, op. cit., p109.
36. Hoffmeyer, *Signs of Meaning*, op. cit., p98.
37. A. Phillips, 'A Stab at Hinting', *The Beast in the Nursery*, London: Faber & Faber, 1998.
38. Polanyi, *Personal Knowledge*, op. cit., p64.
39. Pert, *Molecules of Emotion*, op. cit., p315.
40. J. Hadamard, *The Psychology of Invention in the Mathematical Field*, Princeton: Princeton University Press, Appendix II, titled 'A testimonial from Professor Einstein'.
41. J. Hawkings & S. Blakeslee, *On Intelligence*, Times Books, 2004.
42. Hoffmeyer, *Signs of Meaning*, op. cit., p126.
43. G. Edelman, *Bright Air, Brilliant Fire: On the Matter of Mind*, Harmondsworth, Penguin, 1994.
44. Martin, *The Sickening Mind*, op. cit.; C. Pert, *Molecules of Emotion*, op. cit..
45. Boden, *The Creative Mind*, op. cit., p131.
46. Boden, *The Creative Mind*, ibid., p133.
47. Boden, *The Creative Mind*, ibid., pp133-4.

48. Boden, *The Creative Mind*, ibid., p134.
49. Donald, *Origins of the Modern Mind*, op. cit., p189.
50. Boden, *The Creative Mind*, op. cit., p270.
51. A. MacIntyre, *After Virtue*, op. cit.
52. Laughlin, *A Different Universe*, op. cit.

Bibliography

Abram, D., *The Spell of the Sensuous*, New York: Vintage, 1997.
Ader, R & N. Cohen, 'Psychoneuroimmunology: conditioning and stress, *Annual Review of Psychology*, 44, 1993.
Alexander, V.N., 'Hopeful Monsters: Literary Teleology and Emergence', E:CO vol. 7, Nos. 3 and 4. (Forthcoming 2006)
Anderson, M., J. Deely, M. Krampen, J. Ransdell, T. A. Sebeok, T. von Uexküll, *Semiotica*, vol. 52, 1-2, 1984.
Anderson, P., 'Components of the National Culture', in A. Cockburn & R. Blackburn (eds.), *Student Power: Problems, Diagnosis, Action*, Harmondsworth: Penguin (in association with *New Left Review*), 1969.
Arnold, M., 'The Popular Education of France' (1861), M. Arnold, *Selected Prose*, Penguin: Harmondsworth, 1987.
Attridge, D., *The Singularity of Literature*, London: Routledge, 2004.
Bachen, E., S. Cohen & A.L. Marsland, 'Psychoimmunology', in A. Baum et al, *Cambridge Handbook of Psychology, Health and Medicine*, Cambridge: CUP, 1997.
Ball, P., *Critical Mass: How One Thing Leads to Another*, London: Arrow Books, 2005.
Barabási, A-L., *Linked: The New Science of Networks*, Cambridge, Mass: Perseus, 2002.
Benjamin, W., 'The Storyteller', *Illuminations*, tr. H. Zohn, New York: Schocken Books, 1968.
Berman, M., *All That Is Solid Melts Into Air: The Experience of Modernity*, London: Verso, 1983.
Bertalanffy, L. von, 'The theory of open systems in physics and biology', *Science*, 111: 23-29, 1950.
Bertalanffy, L. von, 'An outline of General Systems Theory', *British Journal for the Philosophy of Science*, 1 139-164, 1950.
Boden, M.A., *The Creative Mind: Myths and Mechanisms*, 2nd ed., London: Routledge, 2004.
Bohm, D., *Wholeness and the Implicate Order*, London: Routledge & Kegan Paul, 1980.
Briggs, J. & F. David Peat, *Turbulent Mirror*, London: Harper & Row, 1989.

Byrne, D., *Complexity Theory and the Social Sciences*, London: Routledge, 1998
Byrne, D., 'The Politics of Complexity: Acting Locally Matters', *Soundings* 14, Spring 2000.
Carroll, J., *Literary Darwinism: Evolution, Human Nature, and Literature*, London: Routledge, 2004.
Cassidy, T., *Stress, Cognition and Health*, London: Routledge, 1999.
Cavalli-Sforza, L.L., *Genes, Peoples and Languages*, tr. M. Seielstad, Harmondsworth: Penguin, 2001.
Cobley, P., (ed.), *The Routledge Companion to Semiotics and Linguistics*, London: Routledge, 2001.
Cohen, S. & G.M. Williamson, 'Stress and Infectious Disease in Humans', *Psychological Bulletin*, 109.
Collier, A., *Critical Realism: An Introduction to Roy Bhaskar's Philosophy*, London: Verso, London, 1994.
Dalal, F., *Race, Colour and the Processes of Racialization: New Perspectives from Group Analysis, Psychoanalysis and Sociology*, London: Brunner-Routledge, 2002.
Damasio, A., *Descartes' Error: Emotion, Reason and the Human Brain*, London: Picador, 1995.
Davis, P., *The Oxford English Literary History, Vol 8. 1830-1880, The Victorians*, Oxford: OUP, 2002.
Dawkins, R., *The Selfish Gene*, Oxford: OUP, 1976.
de Botton, A., *Status Anxiety*, Harmondsworth: Penguin, 2005.
Deely, J., 'The Impact of Semiotics on Philosophy', www.helsinki.fi/science/commens/papers/greenbook.pdf.
Deely, J., *Four Ages of Understanding: The First Postmodern Survey of Philosophy from Ancient Times to the Turn of the Twenty-First Century*, Toronto: Toronto University Press, 2001.
Deely, J., *The Impact on Philosophy of Semiotics*, South Bend, Indiana: St. Augustine's Press, 2003.
Diamond, J., *The Rise and Fall of the Third Chimpanzee*, London: Vintage, 2002.
Donald, M., *Origins of the Modern Mind: Three Stages in the Evolution of Culture and Cognition*, London: Harvard UP, 1991.
Duffy, E., *The Stripping of the Altars: Traditional Religion in England 1400-1580*, New Haven, Conn: Yale U.P, 1992.
Eagleton, T., *After Theory*, London: Allen Lane, 2003.
Edelman, G., *Bright Air, Brilliant Fire: On the Matter of Mind*, Harmondsworth: Penguin, 1994.
Eldredge, N. & S.J. Gould, 'Punctuated Equilibria: an alternative to phyletic gradualism', in T.J.M. Schopf (ed.), *Models in Paleobiology*, San Francisco: Freeman Cooper & Co., 1972.
Eliade, M., *The Sacred and the Profane: The Nature of Religion*, tr. W.R. Trask, London: Harcourt Inc., 1959.
Elias, N., *The Civilizing Process: The History of Manners and State*

Formation and Civilisation, tr. E. Jephcott, Oxford: Blackwell, 1994.
Fox Keller, E., *The Century of the Gene*, London: Harvard University Press, 2000.
Freire, P., *Pedagogy of the Oppressed*, tr. M. Bergman Ramos, New York: Continuum, 2000.
Fudge, E., *Brutal Reasoning: Animals, Rationality and Humanity in Early Modern England*, Ithaca: Cornell University Press, 2006.
Goleman, G., *Emotional Intelligence: Why it can Matter More than IQ*, London: Bloomsbury, 1996.
Goodwin, B., *How The Leopard Changed its Spots: The Evolution of Complexity*, London: Weidenfeld & Nicolson, 1994.
Goody, J., *Capitalism and Modernity: The Great Debate*, London: Polity, 2004.
Gribbin, J., *Deep Simplicity: Chaos, Complexity and the Emergence of Life*, London: Penguin, 2005.
Hadamard, J., *The Psychology of Invention in the Mathematical Field*, Princeton: Princeton University Press, 1945.
Harvey, D., *The Condition of Postmodernity: An Enquiry into the Origins of Cultural Change*, Oxford: Blackwell, 1989.
Hawkings, J. & S. Blakeslee, *On Intelligence*, Times Books, 2004.
Hegel, G.W.F., *Phenomenology of Spirit*, tr. A.V. Miller, Oxford: OUP, 1977.
Hoffmeyer, J., *Signs of Meaning in the Universe*, Bloomington: Indiana University Press, 1996.
Hoffmeyer, J., 'Origin of Species by Natural Translation', www.molbio.ku.dk/MolBioPages/abk/PersonalPages/Jesper/Hoffmeyer.html.
Husserl, E., *The Crisis of European Sciences and Transcendental Phenomenology*, tr. D. Carr, Evanston, Illinois: Northwestern University Press, 1970.
Jablonka, E. and M.J. Lamb, *Evolution in Four Dimensions: Genetic, Epigenetic, Behavioral, and Symbolic Variation in the History of Life*, Cambridge MA: MIT, 2005.
Kauffman, S., *Investigations*, Oxford: OUP, 2000.
Kauffman, S., Foreword to A. Keskinen, M. Aaltonen & E. Mitleton-Kelly, *Organisational Complexity*, FFRC Publications 6/2003, www.tukk.fi/toto/Julkaisut/pdf/Tutu_6_03.pdf.
Koestler, A., *The Act of Creation*, London: Picador, 1975.
Kuhn, T.A., *The Structure of Scientific Revolutions*, London: University of Chicago Press, 1962.
Kull, K., 'Introduction: Entering a semiotic landscape', C. Emmeche, K. Kull, F. Stjernfelt, *Reading Hoffmeyer, rethinking biology*, Tartu: Tartu University Press, 2002.
Kull, K., 'On semiosis, Umwelt, and semiosphere', *Semiotica*, vol. 120 (3/4), 1998. Also available on-line at http://www.zbi.ee/~kalevi/jesphohp.htm.

Kundera, M., 'The Depreciated Legacy of Cervantes', *The Art of the Novel*, London: Faber and Faber, 1988.
Lakoff, G. & G. Fauconnier, *The Way We Think: Conceptual Blending and the Mind's Hidden Complexities*, New York: Basic Books, 2002.
Lakoff, G. & M. Johnson, *Metaphors We Live By*, London: University of Chicago Press, 1980.
Lakoff, G. & M. Johnson, *Philosophy in the Flesh: The Embodied Mind and Its Challenge to Western Thought*, New York: Basic Books, 1999.
Lakoff, G., *Women, Fire, and Dangerous Things: What categories reveal about the Mind*, London: University of Chicago Press, 1987.
Laughlin, R.B., *A Different Universe (Reinventing Physics from the Bottom Down)*, New York: Basic Books, 2005.
Leavis, F.R. & D. Thompson, *Culture and Environment*, London: Chatto & Windus, 1960.
MacIntyre, A., *After Virtue*, London: Duckworth, 1981.
Macy, J., *Mutual Causality in Buddhism and General Systems Theory: the Dharma of Natural Systems*, Albany: SUNY Press, 1991.
Magee, M. & M. Milligan, *Sight Unseen*, Oxford: OUP, 1995.
Margulis, L., *The Symbiotic Planet: A New Look at Evolution*, London: Phoenix, 1999.
Marmot, M., *Status Syndrome: How Your Social Standing Directly Affects Your Health and Life Expectancy*, London: Bloomsbury, 2004.
Martin, P., *The Sickening Mind: Brain, Behaviour, Immunity and Disease*, London: HarperCollins, 1997.
Maturana, H. & F. Varela, *The Tree of Knowledge*, Boston: Shambala, 1987.
Merleau-Ponty, M., 'Science and the Experience of Expression', *The Prose of the World*, Evanston: Northwestern University Press, 1973.
Midgley, M., *Science and Poetry*, London: Routledge, 2001.
Mill, J.S. and J. Bentham, *Utilitarianism and Other Essays*, A. Ryan (ed.), Harmondsworth: Penguin, 1987.
Moran, D., *Introduction to Phenomenology*, London: Routledge, 2000.
Peirce, C.S. 'Design and Chance', *The Essential Peirce: Selected Philosophical Writings*, Vol. 1 (1867-1893), eds. N. Houser & C. Kloesel, Bloomington: Indiana University Press, 1992.
Pert, C., *Molecules of Emotion: Why You Feel the Way You Feel*, London: Pocket Books, 1999.
Phillips, A., 'A Stab at Hinting', *The Beast in the Nursery*, London: Faber & Faber, 1998.
Pickering, J., 'Selfhood as Process', in J. Pickering (ed.), *The Authority of Experience: Essays on Buddhism and Psychology*, Richmond: Curzon, 1997.
Polanyi, K., *The Great Transformation: the political and economic origins of our time*, Boston: Beacon Press, 1957 [1944].

Polanyi, M., *Personal Knowledge: Towards a Post-Critical Philosophy*, London: Routledge & Kegan Paul, 1962 [1958].
Polanyi, M., *The Tacit Dimension*, London: Routledge & Kegan Paul, 1967.
Prigogine, I. & I. Stengers, *Order Out of Chaos: Man's New Dialogue with Nature*, London: Flamingo, 1985.
Prigogine, I., http://www.nobel.se/chemistry/laureates/1977/prigogine-autobio.html.
Putnam, R.D., *Bowling Alone: The Collapse and Revival of American Community*, London: Simon & Schuster, 2000.
Putnam, R.D., 'Introduction', *Democracies in Flux: The Evolution of Social Capital in Contemporary Society*, Oxford: OUP, 2002.
Robertson, J., *After The Welfare State*, 1980, www.jamesrobertson.com.
Rylance, R., *Victorian Psychology and British Culture 1850-1880*, Oxford: OUP, 2000.
Sands, P., *Lawless World: America and the Making and Breaking of Global Rules*, London: Allen Lane, 2005.
Sarafino, E.P., *Health Psychology: Biopsychosocial Interactions*, New York: Wiley, 1998.
Sebeok, T.A. and J. Umiker-Sebeok, *Biosemiotics: the semiotic web*, The Hague: Mouton de Gruyter, 1991.
Sebeok, T.A., *Semiotica*, vol. 52-1/2, 1984.
Sebeok, T.A., 'Nonverbal Communication', in P. Cobley (ed.), *The Routledge Companion to Semiotics and Linguistics*, London: Routledge, 2001.
Sen, A., *Inequality Reexamined*, Oxford: OUP, 1992.
Senghas, A. and M. Coppola, 'Children creating language: How Nicaraguan Sign Language acquired a spatial grammar', *Psychological Science*, 12, 4, 2001.
Senghas, R.J., A. Senghas, and J E. Pyers, 'The emergence of Nicaraguan Sign Language: Questions of development, acquisition, and evolution' in J. Langer, S. T. Parker, & C. Milbrath (eds.), *Biology and Knowledge revisited: From neurogenesis to psychogenesis*, Mahwah, NJ: Lawrence Erlbaum Associates, 2004.
Sennett, R., *The Corrosion of Character: The Personal Consequences of Work in the New Capitalism*, London: W.W. Norton & Co., 1998.
Simpson, D., *On the Origins of Modern Critical Thought: German aesthetic and literary criticism from Lessing to Hegel*, Cambridge: Cambridge University Press, 1988.
Soper, K., 'Postmodernism, Subjectivity and the Question of Value', in J. Squires, (ed.), *Principled Positions: Postmodernism and the Rediscovery of Value*, London: Lawrence & Wishart, 1993.
Soper, K., *What is Nature?*, Oxford: Blackwell, 1995.
Sudnow, D., *Ways of the Hand: A Rewritten Account*, Foreword by Hubert Dreyfus, Cambridge, Mass: MIT Press, 2001.

Taylor, C., *Sources of the Self*, Cambridge: CUP, 1989.
Varela, F.J., E. Thompson, E. Rosch, *The Embodied Mind: Cognitive Science and Human Experience*, London: MIT Press, 1993.
Waldron, J., *God, Locke and Equality: Christian Foundations In Locke's Political Thought*, Cambridge: CUP, 2002.
Wheeler, W., 'Nostalgia Isn't Nasty', in M. Perryman (ed.), *Altered States: Postmodernism, Politics, Culture*, London: Lawrence & Wishart, 1994.
Wheeler, W., 'The Theo-Ontological Expansion of Science', *New Formations*, 50, Summer 2003.
Wheeler, W., *A New Modernity? Change in Science, Literature and Politics*, Lawrence & Wishart, 1999.
Wilkinson, R., *Unhealthy Societies: The Afflictions of Inequality*, London: Routledge, 1996.
Williams, R., *Culture and Society: 1780-1950*, London: Hogarth Press, 1987 [Chatto & Windus, 1958].
Williams, R., *The Long Revolution*, Peterborough, Ontario: Broadview Press Encore Editions, unrevised reprint of London: Chatto & Windus, 1961.
Williams, R., *Keywords*, London: Fontana Press, 1988.
Williams, R., 'Crisis in English Studies', in *The Raymond Williams Reader*, J. Higgins (ed.), Oxford: Blackwell, 2001.
Winnicott, D.W., 'Transitional Objects and Transitional Phenomena', *Playing and Reality*, London: Routledge, 1991.
Wordsworth, W., 1802 Preface to *Lyrical Ballads* in Wordsworth and Coleridge, *Lyrical Ballads*, R.L. Brett and A.R. Jones (eds.), London: Methuen, 1968.

Index

Abram, D., 86, 103n
Ader, R., 129n
Age of Emergence, 12, 20, 25, 28, 30, 32, 145, 155
Age of Reduction, 12, 25, 62, 145
Alexander, V. N., 136, 158n
Anderson, M., 128n
Anderson, P., 16, 57n
Ants and termites, 107, 152
Arnold, M., 84-85
Art, 16, 33, 44, 47, 49-51, 55, 58n, 63, 69, 81, 83, 90, 93, 98, 106, 131-132, 135, 137
Attridge, D., 134, 158n
Autopoiesis, autopoietic, 41, 46, 53-54, 73, 94, 107, 136, 139, 141, 151

Bachen, E., 129n
Ball, P., 138, 158n
Barabási, A-L, 31, 37n
Benjamin, W., 85, 103n
Bentham, J., 27, 37n, 82
Berman, M., 132, 158n
Bertalanffy, L. von, 52, 58n, 98
Bhaskar, Roy, 54-55, 66, 73
Biology, 13-17, 19, 22, 24-25, 27-29, 31, 33-34, 37n, 43, 48, 52, 53, 58n, 60, 66, 70-75, 101-102, 110-112, 120-121, 127, 128n, 129n, 132-134, 138-139, 155, 158, 159n; of complex systems, 19, 33, 48, 52-53, 60, 72, 155; and culture, 22, 75, 110-112, 132, 134, 139, 155
Biosemiotic, biosemiotics, biosemiosis, 19, 24, 34, 56, 107-108, 110-112, 114-115, 117, 120, 122-124, 126-127, 132-135, 138-139, 141-143, 146, 155-157
Blakeslee, S., 159n
Boden, M., 34, 37n, 70, 78n, 81, 139-141, 151-155, 159n, 160n
Bohm, D., 67-68, 78n
Briggs, J., 53, 58n
Buddhism, Buddhist, 58n, 81, 87-88, 92, 94, 97, 99-100, 104n
Byrne, D., 46, 57n, 58n

Carroll, J., 26, 36n
Cassidy, T., 128n
Cavalli-Sforza, L.L., 33, 37n
Cell, 13, 28-29, 53, 71, 73, 107, 117, 120-122, 124-125, 133, 145, 151
Chance, 136, 147, 149, 154, 158n, 159; see also under coincidence and serendipity
Class, 30, 45, 51, 60, 101, 112, 126
Cobley, P., n36, 128n
Cohen, N., 129n
Cohen, S., 129n
Coincidence, 136; see also under chance and serendipity

167

Coleridge, S.T., 27, 42, 103n, 109
Collier, A., 58n
Communication, 13-14, 16-17, 26, 56, 68, 99, 110, 112, 117-118, 121, 123, 128n, 131, 132, 135, 140, 143, 144, 153; in animals, 143; in plants; 121; see also semiotics, semiosis, etc
Complexity, 13, 14-16, 18-21, 25, 28, 33, 33, 34, 39-41, 43-44, 46-48, 51-56, 57n, 58n, 60, 66-70, 72, 74-75, 77n, 81-82, 84, 86, 92-94, 97, 101, 103, 106, 108, 112, 114, 117, 121, 125-126, 128n, 132, 136-137, 148, 152-153, 155, 159n
Computers, 52, 151, 152
Conservation, 102, 156
Creative, creativity, 15, 19, 21, 23, 29, 32-35, 37n, 40, 44-45, 47, 50, 51, 52-53, 63, 67, 69-70, 73, 75-76, 78n, 81-82, 89-91, 93, 102, 113-114, 125, 127, 131-143, 146-156, 159n; ancient and iconic, 137; as autopoiesis, 73; and autopoietic semiosis, 139; and biosemiotic freedom, 132, 153; and code-duality, 139; as combinational, 139; and communication, 132, 140; and complex systems, 155; and computers, 151; and creative ethos, 135; and culture and language, 134; and environment and culture, 142; and frame-breaking, 70, 81, 125, 136; and human evolution, 143; and inequality, 113, 114; and labour, 137; as manipulation of scripts and frames and DNA, 139; and markets, 141; and science, 90, 91; and semiotic liveliness, 139; as signature of life, 133; and symbiogenesis in culture, 134; and tacit knowledge, 67; and time, 141; and Time's Arrow, 141; as transformational, 140
Culture, 13, 14, 15, 17, 16, 19, 21, 22, 25, 26, 27, 28, 32, 34, 35, 36n, 37n, 41, 44, 45, 57n, 60, 67, 78n, 83, 86, 90, 93, 94, 95, 98, 102, 103n, 108, 109, 113, 114, 123, 124, 125, 126, 127, 131, 132, 134, 135, 137, 138, 139, 140, 141, 142, 144, 145, 146, 147, 150, 153, 154, 155, 157, 158, 158n; and biology and nature, 22, 25, 26, 28, 32, 34, 41, 44, 86; and ethics, 35; and evolution, 44, 67, 114, 142
Culture and Society: 1780-1950, 15, 16, 19

Dalal, F., 54, 58n
Damasio, A., 58n, 82, 103n
Davis, P., 84-85, 103n
Dawkins, R., 58n, 78n
De Botton, A., 113, 128n
Deely, J., 16-17, 24, 35n, 36n, 81, 121, 128n
Diamond, J., 36n
DNA, 14, 29, 71, 73-74, 117, 123-125, 133, 139-140
Donald, M., 33, 37n, 78n, 142-143, 145, 154, 159n, 160n
Duffy, E., 57n

Eagleton, T., 39, 101, 105n
Edelman, G., 159n
Eldredge, N., 77n
Eliade, M., 95, 96, 97, 104n
Elias, N., 39, 56n, 57n, 69, 96

Emergence, emergent phenomena, 12-13, 19-20, 22, 25-32, 34, 37n, 43, 46, 48, 52, 54-56, 60, 61, 67, 69-75, 87, 90, 93, 98-9, 103, 106-108, 110, 114, 115, 124, 125-127, 128n, 131-132, 134, 136-137, 141, 142, 143-145, 155, 158n, 159n; see also, Age of Emergence

Endocrine system, 32, 108, 116-117, 120, 122, 151

Epidemiology, 17, 31, 80, 108, 110, 145

Epigenetic, 14, 35n

Ethics/ethical/ethos, 14, 35, 111, 134, 135, 136, 137, 154, 156, 157; ethos of responsiveness, 156

Evolution, 13-15, 18-19, 21-25, 28, 31, 33-35, 35n, 36n, 37n, 38, 44, 49, 52-53, 56, 56n, 64, 66-67, 69, 70-72, 75, 77n, 78n, 89, 95, 106-108, 121, 123-126, 129n, 132-134, 136-145, 155-156, 158n, 159n; see also evolution, cultural

Evolution, cultural, 18-19, 37n, 69, 78n, 126, 132, 137, 139, 141, 155; of language, 19, 158n

Fauconnier, G., 140, 159n
Fibonacci numbers, 72, 136
Fox Keller, E., 37n, 123
Frame, frames, 21, 71, 74, 114, 132, 136, 154, 156; frame-breaking, 70, 136, 154
Freud, Sigmund, 54, 92, 99, 138, 146
Freire, P., 78n
Fudge, E., 49, 57n

General Systems Theory, 33, 39, 52, 58n, 94
Genetic reductionism, see reductionism
Goleman, G., 59n
Goodwin, B., 33, 46, 48, 51, 56n, 57n, 58n, 60, 67, 71-75, 78n, 79n, 82, 89, 100, 104n, 106-107, 120, 126, 128n, 149
Goody, J., 36n, 118, 129n
Gould, S.J., 77n
Gribbin, J., 128n, 159n

Hadamard, J., 150, 159n
Harvey, D., 141, 159n
Hawkings, J., 159n
Health/illness, 12, 18, 32, 34, 40, 70, 80, 100, 108, 111-120, 126, 128n, 129n, 148, 152, social basis of, 31, 32,34, 37n, 54, 108, 111-120, 124, 126
Hegel, G.W.F., 25, 36n, 42, 45, 57n, 145
Hoffmeyer, Jesper, 19, 24, 27-28, 32, 32-35, 36n, 37n, 63-64, 71, 74-75, 77n, 78n, 79n, 80, 98, 102-103, 103n, 106-110, 117, 122-123, 125-127, 127n, 128n, 129n, 130n, 132, 133, 134, 142-145, 155, 158n, 159n
Holism/holistic, 21, 26, 45-46, 52, 75, 87
Homo erectus, 142, 144, 154
Homo habilis, 143
Homo sapiens, 19, 22, 28, 110, 143, 144
Husserl, E., 44, 49, 57n, 88, 89, 94

Icon/iconic, 136-137, 144, 147
Imagination, 45, 122, 158
Immune system, 24, 108, 116-

118, 120, 122, 127, 142, 151; immunology/ psychoneuroimmunology, 116
Index, 115, 136
Intersubjective, intersubjectivity, 34, 48, 87, 91, 97, 134, 145, 155-156
Intimation, 65, 68, 90
Intuition/intuitive, 68, 109, 135

Jablonka, Eva, 14
Johnson, M., 140, 159n

Kauffman, S., 29, 37n, 41, 56n
Keats, John, 42, 146-147
Klein, Melanie, 54
Koestler, A., 104n
Krampen, N., 121, 128n
Kuhn, T.A., 19-20, 36n
Kull, K., 102, 105n, 110, 125, 128n, 129n, 130n
Kundera, M., 49, 58n

Lakoff, G., 70, 78n, 86, 104n, 140, 159n
Lamb, Marion J., 14
Language, linguistic, 16-28, 33-34, 39, 43, 45, 47, 55, 61-62, 66, 73, 82, 87, 89, 91, 95, 100-102, 105n, 107-111, 117, 120-127, 132, 134, 136, 139, 140-147, 153-54, 157, 159n; and evolution, 14, 19, 158n
Laughlin, R.B., 20, 25, 28-30, 36n, 37n, 51, 58n, 62, 81, 103n, 104n, 127n, 145, 155, 157, 160n
Leavis, F.R., 158n
Liberal/Liberalism, 17-19, 21, 24, 30, 33-34, 37n, 39, 96, 111-112, 127, 140, 153, 156-158; and freedom, 18, 30; and human sociality, 30; and individualism, 21, 26, 33, 39, 58n, 111, 127, 140; neo-liberal, 18, 26; and reason, 18, 157
Literature, 36n, 48-49, 77, 83, 101-102, 103n, 105n, 158n
Long Revolution, The, 12-16, 25, 35n, 39, 128n, 129n, 131, 137, 158n

MacIntyre, A., 111, 128n, 156, 160n
Macy, J., 52, 58n, 94, 97, 100, 104n, 105n
Magee, B., 56n, 78n
Margulis, L., 35, 121, 133, 158n
Marmot, M., 32, 35, 37n, 76, 82, 108-118, 120, 126, 128n, 148, 152
Marsland, A.L., 129n
Martin, P., 106, 108, 116, 117, 118, 120, 129n, 148
Maturana, H., 24, 36n, 48, 106-107, 120, 128n
Memory, 83, 107, 121, 122, 123, 124, 147; swarm memory, 122
Merleau-Ponty, M., 87, 104n
Message/messages, 23-24, 32, 107, 110, 117, 121-122, 125-126, 134, 143, 145-146, 151; see also, semiotics, communication
Midgley, M., 56n, 57n
Mill, J.S., 37n, 82
Mimetic/mimeticism/mimesis, 142-146, 153-154
Modernity, 13, 15, 17, 36n, 39-42, 49, 51, 76, 86, 91, 96-97, 103n, 118, 129n, 131-132, 138, 140-141, 155-156, 158n
Moran, D., 57n

Index

mRNA 29, 71, 117, 139
Muse, 146

Nature, 12-16, 19, 21-27, 30-34, 36n, 40-41, 44-45, 50, 53-54, 56, 58n, 60-67, 71-75, 81, 83, 86, 88, 90-99, 101-102, 104n, 106, 109, 117, 122-123, 125, 127, 131-132, 134-135, 137-140, 148-150, 153, 155-157
Nervous system, 32, 53, 108, 116-117, 120, 122, 151
Networks, 37n, 119, 146, 150; associative networks, 146; neural networks, 150; social networks, 119
Neuropeptides, 82, 116-117, 148
Nicaraguan Sign Language, 33, 37n, 126, 142, 144, 159n
Non-linearity/non-linear, 22, 39, 41, 43, 46, 53-54, 125, 140

Paradigm shift, 12-13, 20-21
Parallel processing, 151-152
Passion, 44, 50, 55, 68-69, 76, 80, 86, 148
Peat, F.D., 53, 58n
Peckham Experiment, 46
Peirce, Charles Sanders, 16, 17, 32, 52, 64, 103, 120, 121,158, 158n
Period-doubling, 141
Pert, C., 20, 36n, 82, 90, 104n, 117, 129n, 148, 151, 159n
Phillips, A., 147, 149, 159n
Pickering, J., 86, 87, 88, 94, 104n
Poet/poetry, 44, 56n, 57n, 60, 84-85, 139-140, 146-147
Polanyi, K., 35, 58n
Polanyi, M., 25, 33, 42, 48-50, 55-56, 58n, 60-68, 70, 72, 74, 76-77, 77n, 78n, 79n, 81-82, 86, 89-90, 93, 98-99, 104n, 108, 125, 131, 133, 135, 137, 146, 147, 149, 158n, 159n
Postmodernism, postmodernity, 26, 36n, 38, 101, 103n, 140, 159n
Prigogine, I., 51, 52, 54, 58n, 93, 121
Psychoanalysis, 40, 54, 58n, 97, 153
Putnam, R.D., 79n, 106, 113, 119, 120, 128n, 129n

Ransdell, J., 128n
Reason, 16, 18, 21, 33, 41-42, 58n, 75, 80, 83, 91, 103n, 113-114, 138, 149, 150, 157-158, 159n
Reduction, 12-13, 22, 32, 4- 43, 68, 134, 139, 145, 155; of other to the same, 134; see also, Age of Reduction
Reductionism, 13, 17, 20-22, 25-26, 29-31, 43, 55, 62, 139; genetic, 22
Religion, 30, 33, 42-43, 49, 57n, 76, 80-86, 88-98, 104n, 106, 137, 140
Revolution, 12, 15, 19, 132; complexity revolution, 20, 38, 43, 69; industrial revolution, 83; scientific revolution, 42, 69, 71, 83; and tacit knowledge, 61
Robertson, J., 137, 158n
Romantic/romanticism, 18, 21, 24-25, 30, 33, 42-45, 55, 57n, 70, 86, 101, 127
Rosch, E., 87, 104n
Rules, 19, 36n, 52, 89, 93, 107, 125, 128, 136, 138-140, 146-147, 151; rule-breaking, 19, 147-148

Rylance, R., 20, 36n

Sands, P., 26, 36n
Sarafino, E.P., 128n
Science, 13, 16-21, 25-34, 36n, 37n, 38-56, 56n, 57n, 58n, 60, 62-64, 66, 68-75, 80-83, 86-87, 88, 90-102, 103n, 104n, 106, 108, 111-112, 118, 124, 128n, 132, 147-149, 153-154, 159n; as cultural evolution, 18, 19; and discourse, 19; and liberalism, 19; and myth, 71; and passion, 50, 55, 68, 69, 76, 86, 148; from reduction to emergence, 25
Science of qualities, 51, 70, 74, 82, 149
Sebeok, T.A., 16, 27-28, 34, 37n, 81, 102-103, 110-111, 120-121, 128n, 132, 143
Self-organisation, 46, 53, 141, 151, 155
Semiosis, 16, 17, 19, 24, 27, 28, 32, 34, 47, 48, 63, 87, 102-103, 105n, 106-111, 117-118, 120, 122-123, 126, 132, 137, 143, 146-147, 151, 155-156
Semiotic freedom, 18, 19, 24, 28, 75, 95, 98, 107, 109, 112, 114, 124-127, 133, 138, 147, 152-155
Semiotic/semiotics, 14-19, 21, 24, 27-30, 34, 35n, 36n, 37, 47-48, 56, 63, 66, 74-75, 95, 98-99, 101-103, 107-112, 114, 120, 123-127, 128n, 129n, 130n, 132-133, 138-139, 143, 145, 147, 150, 152-157; psychosemiotic, 99; semiosymbiogenesis, 134; semiotic web 37; see also communication

Sen, A., 98, 128n
Senghas, A., 37n, 159n
Sennett, R., 36n, 79n
Serendipity 90, 149, 154; see also chance and coincidence
Simpson, D., 57n
Socialism/socialist, 30, 35, 40-41, 44-46, 57n, 60, 112, 132, 155-157
Sociality, 12-13, 21-22, 27, 30-31, 34, 76, 82, 99, 106, 110-111, 115-116, 118-119, 132, 153
Society, 21, 25-26, 28-29, 32, 35, 38, 39, 41-46, 54, 57n, 58n, 60, 69, 70, 76, 82, 10- 109, 114, 118, 120, 126-127, 129n, 131-132, 134, 137-139, 142, 156; and class 44, 45, 60; as complex and evolutionary, 38-40, 45, 114; creative society, 35; good society, 21; no such thing as, 26
Soper, K., 18, 36n, 86, 92, 104n
Status, 32, 35, 37n, 108, 111-115, 126, 128n, 142, 148
Stengers, I., 54, 58n
Stratification/Strata, 22, 32, 43, 54-56, 60-61, 66- 67, 73, 90, 106, 108, 110, 127, 133, 139, 145, 155
Sudnow, D., 62, 77n, 89, 104n
Swarm intelligence, 32, 35, 107, 123, 146, 150-152
Swarm of swarms, 117
Symbiogenetic, symbiogenesis, 13, 133-134, 137; as failed eating, 134; as semiosymbiogenesis, 134

Tacit knowledge, 25, 28, 33, 42, 44, 47-51, 54-56, 56n, 60-69, 77, 77n, 78n, 79n, 87, 89-91, 97, 99-100, 106, 108, 132, 135-

137, 139, 142, 145-148, 150, 158n; in complex systems, 25, 54, 60-63, 106, 108
Taylor, C., 111, 128n
Teleology, teleological, 136, 158n
Thompson, D., 258n
Thompson, E., 87, 104n, 158n

Uexküll, J. von, 34, 102, 103, 120
Uexküll, T. von, 128n
Umwelt, 23-24, 28, 34, 103, 105n, 107-108, 110, 120-122, 124-126, 133-134, 137-138, 140-141, 143, 145, 148, 150, 155
Utilitarian, Utilitarianism, 21, 44, 80, 113

Varela, F., 24, 36n, 48, 87, 104n, 106, 107, 120, 128n

Waldron, J., 37n
Wheeler, Tilly, 148
Wheeler, W., 86, 103n
Wilkinson, R., 58n, 110
Williams, R., 12-13, 14-15, 35n, 36n, 47, 57n, 86, 104n, 109, 128n, 129n, 131-132, 137, 158n, 166n; *Long Revolution, The*, 12-16, 25, 35n, 39, 128n, 129n, 131, 137, 158n; *Culture and Society: 1780-1950*, 15, 16, 19
Williamson, G.M., 129n
Winnicott, D.W., 158n
Wordsworth, William, 42, 60, 82, 85, 103n
Work, 116, 118, 126, 127, 135, 157; good work, 135, 150; pre-political work in culture, 137, 138; work-related stress, 118
World-modelling, 106